DISABILITY IN THE INDUSTRIAL REVOLUTION

Manchester University Press

Series editors
Dr Julie Anderson, Professor Walton Schalick, III

This new series published by Manchester University Press responds to the growing interest in disability as a discipline worthy of historical research. The series has a broad international historical remit, encompassing issues that include class, race, gender, age, war, medical treatment, professionalisation, environments, work, institutions and cultural and social aspects of disablement including representations of disabled people in literature, film, art and the media.

Already published
Deafness, community and culture in Britain: leisure and cohesion, 1945–1995
Martin Atherton
Rethinking modern prostheses in Anglo-American commodity cultures, 1820–1939
Claire L. Jones (ed.)
Destigmatising mental illness? Professional politics and public education in Britain, 1870–1970
Vicky Long
Intellectual disability: a conceptual history, 1200–1900
Patrick McDonagh, C. F. Goodey and Tim Stainton (eds)
Fools and idiots? Intellectual disability in the Middle Ages
Irina Metzler
Framing the moron: the social construction of feeble-mindedness in the American eugenics era
Gerald V. O'Brien
Recycling the disabled: army, medicine, and modernity in WWI Germany
Heather R. Perry
Worth saving: disabled children during the Second World War
Sue Wheatcroft

DISABILITY IN THE INDUSTRIAL REVOLUTION

PHYSICAL IMPAIRMENT IN BRITISH COALMINING, 1780–1880

David M. Turner and Daniel Blackie

Manchester University Press

Copyright © David M. Turner and Daniel Blackie 2018

The rights of David M. Turner and Daniel Blackie to be identified as the authors of this work have been asserted by them in accordance with the Copyright, Designs and Patents Act 1988.

Published by Manchester University Press
Altrincham Street, Manchester M1 7JA

www.manchesteruniversitypress.co.uk

British Library Cataloguing-in-Publication Data
A catalogue record for this book is available from the British Library

ISBN   978 1 5261 1815 8 hardback
ISBN   978 1 5261 2577 4 open access

An electronic version of this book is also available under a Creative Commons (CC-BY-NC-ND) licence, thanks to the support of the Wellcome Trust

First published 2018

The publisher has no responsibility for the persistence or accuracy of URLs for any external or third-party internet websites referred to in this book, and does not guarantee that any content on such websites is, or will remain, accurate or appropriate.

Typeset in 10/12pt Arno Pro by
Servis Filmsetting Ltd, Stockport, Cheshire
Printed by Lightning Source

In memory of Anne Borsay

Contents

List of figures viii
Series editors' foreword ix
Acknowledgements x
List of abbreviations xii

Introduction 1
1 Disability and work in the coal economy 23
2 Medicine and the miner's body 55
3 Disability and welfare 93
4 Disability, family and community 128
5 The industrial politics of disablement 163
Conclusion 200

Select bibliography 208
Index 223

Figures

1. British coalfields in the nineteenth century, adapted from R. A. S. Redmayne, 'The Coal-Mining Industry of the United Kingdom', *The Engineering Magazine*, xxvi (1904). Credit: Notuncurious/ Wikimedia Commons/CC-BY-SA-3.0. — 10
2. A coal mine: miners at work above and below ground. Wellcome Library, London/CC-BY 4.0. — 43
3. 'The Strike in South Wales: Interior of a Collier's Cottage', *Illustrated London News*, 18 January 1873. Copyright Illustrated London News Ltd/Mary Evans. — 137
4. 'Capital and Labour', *Punch*, 29 July 1843. Reproduced with the permission of Punch Ltd. Punch.co.uk. — 164
5. 'Pitmen Encamped', *Illustrated London News*, 3 August 1844. Copyright Illustrated London News Ltd/Mary Evans. — 181

Series editors' foreword

You know a subject has achieved maturity when a book series is dedicated to it. In the case of disability, while it has co-existed with human beings for centuries the study of disability's history is still quite young.

In setting up this series, we chose to encourage multi-methodologic history rather than a purely traditional historical approach, as researchers in disability history come from a wide variety of disciplinary backgrounds. Equally 'disability' history is a diverse topic which benefits from a variety of approaches in order to appreciate its multi-dimensional characteristics.

A test for the team of authors and editors who bring you this series is typical of most series, but disability also brings other consequential challenges. At this time disability is highly contested as a social category in both developing and developed contexts. Inclusion, philosophy, money, education, visibility, sexuality, identity and exclusion are but a handful of the social categories in play. With this degree of politicisation, language is necessarily a cardinal focus.

In an effort to support the plurality of historical voices, the editors have elected to give fair rein to language. Language is historically contingent and can appear offensive to our contemporary sensitivities. The authors and editors believe that the use of terminology that accurately reflects the historical period of any book in the series will assist readers in their understanding of the history of disability in time and place.

Finally, disability offers the cultural, social and intellectual historian a new 'take' on the world we know. We see disability history as one of a few nascent fields with the potential to reposition our understanding of the flow of cultures, society, institutions, ideas and lived experience. Conceptualisations of 'society' since the early modern period have heavily stressed principles of autonomy, rationality and the subjectivity of the individual agent. Consequently we are frequently oblivious to the historical contingency of the present with respect to those elements. Disability disturbs those foundational features of 'the modern'. Studying disability history helps us resituate our policies, our beliefs and our experiences.

Julie Anderson
Walton O. Schalick, III

Acknowledgements

This book has been written as part of the Wellcome Trust Programme Award in Medical History, 'Disability and Industrial Society: A Comparative Cultural History of British Coalfields, 1780–1948' [grant number 095948/Z/11/Z]. It draws on the work of the research team: Professor Anne Borsay, Professor David Turner, Professor Kirsti Bohata, Dr Mike Mantin and Dr Alexandra Jones (Swansea University); Dr Daniel Blackie (University of Oulu); Dr Steven Thompson and Dr Ben Curtis (Aberystwyth University); Dr Vicky Long (Glasgow Caledonian University) and Dr Victoria Brown (Northumbria University/Glasgow Caledonian University); and Professor Arthur McIvor and Dr Angela Turner (Strathclyde University).

We thank the Wellcome Trust for its generous support of this project. In addition to the collegial and inspiring support from the research team, we are also grateful for the encouragement we have received from members of the project's Advisory Board and Public Engagement Panel, and the many archivists, librarians and curators who have helped us obtain the sources we use in this book. Swansea University's Department of History and Classics provided a friendly and stimulating environment in which to develop our ideas and we thank all our colleagues there for making it such a lovely place to work. The staff of the university's Research Institute for Arts and Humanities provided much appreciated support and helped us greatly in our enjoyable and successful public engagement activities. The College of Arts and Humanities granted David a period of sabbatical leave to work on this book. Thanks also to the many scholars who have helped us sharpen our analysis with their comments at conferences and seminars over the last few years and have made participating in those meetings such a lot of fun.

David would like to thank John Spurr, Elaine Canning and colleagues in the College of Arts and Humanities, and staff at the Wellcome Trust for supporting him as project leader, particularly following the death of Anne Borsay in 2014. Angela John, Alun Withey and Andy Croll have all provided much needed encouragement and shared the insights of their own research. Conversations with the community of disability historians and students at Swansea University have helped to shape the ideas of this book further. He is grateful to Lesley Hulonce, Patricia Skinner, Irina Metzler, Gemma Almond, Teresa Hillier, Pallavi Podapati and Rachel Wilks for many stimulating discussions about disability and difference. The love of his wife Carys and son Dyfan, has sustained him through the process of writing this book and he thanks them for their patience and support.

Daniel thanks his family and old friends in Britain for helping him reintegrate into British society after years away experiencing the delights of the North. His new friends in south Wales have been incredibly important too, offering much needed and appreciated friendship, song, and the occasional lift to the gorgeous and sanity-saving Gower Peninsula. The final phase of Daniel's work on this project was completed at the University of Oulu and he is grateful for all the support he has received there. He would particularly like to thank his colleagues in the History of Science and Ideas and History programmes, and the esteemed ladies and gentlemen of the 'Frost Club' for making him feel so welcome. Petteri Pietikäinen, Annukka Sailo and Sami Lakomäki have been especially brilliant in this regard and Daniel is thankful for their help, kindness, and company. Last, but definitely not least, Daniel thanks Katja Huumo for all her love and support over the years. Without her, he admits, he would be a lot more grumpy.

Abbreviations

PP	UK Parliamentary Papers (Proquest), http://parlipapers.proquest.com/
Statistical Compendium	David Turner, Steven Thompson, Kirsti Bohata, Vicky Long, Arthur McIvor, Mike Mantin, Daniel Blackie, Ben Curtis, Angela Turner, Victoria Brown, Alexandra Jones, Anne Borsay, *Disability and Industrial Society, 1780–1948: A Comparative Cultural History of British Coalfields: Statistical Compendium*, http://doi.org/10.5281/zenodo.183686
TNA	The National Archives

INTRODUCTION

In November 1792 there was an explosion of gas at Benwell Colliery, Newcastle-upon-Tyne. Among the victims was James Jackson, a thirty-six-year-old miner, who suffered significant injuries to his face, neck, part of his breast, hands and arms. Burns to his lips and nostrils indicated that he had suffered some internal injuries. When rescuers found him he was shivering, which suggested, in the words of Edward Kentish, the surgeon who treated him, that he had suffered a 'violent shock to the general system'. In the weeks following the accident, Jackson underwent a lengthy and uncomfortable course of treatment. His hands were washed with 'heated essence of turpentine', before being covered with plasters. He was given laudanum for the pain and a teacup full of 'oily emulsion, with an ounce of camphorated tincture of opium' every three hours. His injuries required round-the-clock attention, with bandages applied and reapplied and emollient rubbed on his burnt parts, but at length he began to recover. The skin started to return to his face and hands after a fortnight, and within six weeks Jackson was deemed 'capable of work'. Kentish recorded with pride that his treatment plan had 'combined everything I had to wish: it saved life, it eased pain, and it speedily restored my patient to health and usefulness'. And so Jackson was able to return to work, albeit with a body likely permanently scarred with physical reminders of the dangers of his occupation.[1]

Jackson was a survivor, but many victims of mining accidents were not so fortunate. Fatal accidents, such as the large-scale disasters that claimed 204 lives at Hartley Colliery in Northumberland in 1862 or, worst of all, the explosion that killed 439 men and boys at Universal Colliery, Senghenydd, in 1913 are well known.[2] But as John Benson has pointed out, many British miners were killed in smaller accidents that claimed one or two lives. Still more suffered non-fatal injuries, or contracted chronic diseases that sapped their strength and shortened their lives.[3] Dr James Mitchell, presenting evidence in 1842 to a commission set up to examine the employment of children in coal mines, documented a series of accidents at several unnamed Durham collieries. They included a worker who had lost his leg 'in consequence of coal falling upon it' and one who 'got two fingers taken off by the waggons jamming his hand', leaving him 'maimed'. Another worker, hurt by falling under a horse, 'was five months off work and remains weakly'. The accident had left him '*a little distorted*, but not so as to impede him from working'.[4] As the geologist Henry de la Beche told a House of Lords committee in 1849, although such accidents were 'very considerable', they did not 'excite the notice which is occasioned by explosions in the larger collieries'. 'A great many are occasionally disabled who are never heard of,' he noted, and were subsequently forced into dependency on poor relief 'in consequence of injuries that no one ever hears of.'[5]

This book examines the lives and experiences of these people, men like James Jackson, who, until recently, were 'never heard of' in histories of industrialisation – the scarred, the mutilated, the 'distorted' and the impaired. The process of industrial growth in Britain after 1700, which gathered pace from the late eighteenth century, orchestrated changes in professional, family and community, political and cultural life as well as in the economy and technology. Since the late 1960s, such processes have been examined via perspectives ranging from business history to gender history. Yet disability history is absent from this intellectual endeavour.[6] As we show in the pages that follow, disability was central to the Industrial Revolution. Worries about disability and what to do about the seemingly countless numbers of workers injured in the service of industry prompted policy innovations that continue to affect the lives of Britons today, such as workplace health and safety regulations; age restrictions on when people can start work; and medical institutions catering for specific populations. Not only did disability become visible in its modern forms during the period, it also helped nineteenth-century Britons make sense of the momentous changes happening around them. The existence and experiences of chronically ill or maimed workers were regarded by many as proof of the evils of industrialism, providing a rallying call for the nascent labour movement and a rationale for worker-led campaigns and mutualism that fed their developing class consciousness. Disabled people, as we shall see, contributed to Britain's industrial development, while disability in turn shaped responses to industrialisation.

Given the largely forgotten significance of disability in the Industrial Revolution, what happens to our view of industrialisation when we place people with impairments at the heart of the story? As the examples above suggest, experiences of injured workers resist straightforward generalisation.

For those who became reliant on public welfare after becoming 'disabled', there were others for whom bodily impairment did not necessarily mean an end to their working lives. How did industrial expansion contribute to the incidence of injury, disease and impairment? What happened to those 'disabled' through accidents or disease during Britain's Industrial Revolution? How did people with impairments negotiate changing welfare and medical regimes of assistance, and what was the place of disability in industrial politics? Did industrial change lead to increasing marginalisation of 'disabled' people and how receptive was the workplace to men, women and children with impairments? And what does a study of the Industrial Revolution that foregrounds the experience of disabled people contribute to our understanding of work and its politics in the past?

This book attempts to answer these questions by examining perceptions and experiences of disability within the context of the British coal industry and Britons' responses to people in mining areas who today might be labelled 'disabled'. Coal provides a compelling case study for exploring occupational impairment in industrialising Britain. Coal was at the forefront of the Industrial Revolution, powering, for instance, the expansion of the metallurgical and manufacturing sectors.[7] One of the largest employers of labour, moreover, the industry was one of the most dangerous to work in and mineworkers were exposed to a variety of hazards ranging from noxious and flammable gases to dust, rock falls and equipment failure. Not only were miners at greater risk than any other workers to fatal accidents, they were also at significant risk of injury or disablement, with perhaps 100 non-fatal accidents for every fatal one.[8] In Benson's estimation, during the second half of the nineteenth century, 'a miner was killed every six hours, seriously injured every two hours, and injured badly enough to need a week off work every two or three minutes'.[9] In contrast to histories of disability that explore the 'otherness' of physical difference, such as studies of disabled people's work as 'freak' show performers, this book explores the history of disability within communities where some degree of bodily damage was the norm rather than the exception, where injuries, diseases and ailments were accepted as daily occurrences.[10]

We examine responses to and experiences of disability in a formative period of industrial expansion – the so-called 'classical' phase of the Industrial Revolution. These responses and experiences, as we will see, played out and were shaped in coalfield communities that celebrated social solidarity on the one hand and individual self-reliance on the other. Beginning in 1780, just before the expansion of the Great Northern Coalfield in north-east England, the book addresses the processes of industrialisation related to coalmining and their implications for conceptions and experiences of disability. It sheds light on the various community, political, medical and welfare responses to workers' disability in the century before the 1880 Employers' Liability Act – a landmark, if flawed, legislative intervention that enshrined in law employer responsibility for workplace accidents that could have been prevented.[11] The book therefore charts a shift from ad hoc responses to disability to the first signs of a more formal recognition of the needs of disabled workers in a period that is significant for the gradual evolution of 'disability' as a category distinct from other forms of disease or ill health.[12] It examines the role of economic changes in shaping understandings and experiences of disability during this crucial era of industrial development. Different communal, welfare and medical responses to disablement are analysed alongside evidence that indicates the agency of people with impairments. Indeed, rather than seeing 'disabled' mineworkers simply as the victims of exploitative economic expansion, a key contention of this book is that these people made important contributions to Britain's Industrial Revolution. 'Disabled' people were a conspicuous presence in industrialising Britain, in the workplace and as participants in the community life and industrial politics of Britain's coalfields. The remainder of this introductory chapter sets out the aims and objectives of the book in more detail and provides the historical and methodological context for the discussion that follows.

Disability and industrialisation

If disability has been largely absent from conventional histories of industrialisation, the Industrial Revolution has assumed great significance in disability studies. The idea that industrial economic development has had a profound impact on modern Western understandings and experiences of disability is a pervasive one in the field. Scholars influenced by historical materialism have been at the forefront of this kind of theorising. Writing in the 1980s, Vic Finkelstein provided one of the clearest and boldest statements of this position when he argued that 'disability' is essentially a creation of industrial capitalism. For him, the economic changes of the Industrial Revolution marked a decisive shift in the status of people with impairments during which they found it increasingly difficult to sell their labour on the same terms as others, leading to their increasing stigmatisation and isolation. This theory has been taken up and developed further by other scholars, most notably Michael Oliver and Colin Barnes, and Brendan Gleeson.[13]

Prior to industrialisation, it is argued, physically impaired people may have experienced poverty and stigma, but the organisation of society was such that it enabled them to participate in daily life to the best of their abilities.

The predominantly agrarian nature of the pre-industrial economy, where production centred on the home and workers worked to task, meant that people had greater autonomy to decide their own work routines, rhythms and practices. Although impairment might prove challenging, then, the structure, requirements and expectations of pre-industrial life were flexible enough to allow permanently injured or chronically ill people to take up productive or other socially valued roles. With the coming of industrialisation, however, this 'somatic flexibility', as Gleeson terms it, was significantly undermined and impaired people were forced into less socially desirable positions.[14]

Building on Finkelstein and other materialist accounts, Oliver and Barnes point to four key 'disabling' elements of industrial societies: the growing speed of production associated with mechanised factory work; stricter discipline of workforces; more stringent time keeping; and the standardisation and regulation of production norms. Together, these are believed to have made workplaces hostile and unaccommodating environments for people with impairments. If they were not excluded from work altogether, impaired people were, at best, relegated to marginal productive roles that were poorly rewarded and of low status. As a result, people with impairments became 'disabled', stigmatised as unproductive and pushed to the margins of society. Increasingly regarded as a problem, disabled people in industrial societies were subjected to institutional 'solutions' that saw many placed in specially created facilities and segregated from the wider community. This belief in the institutionalising impulse of industrial societies was expressed most forcefully perhaps by Finkelstein. But others also maintain the premise, albeit in a slightly modified form. By the end of industrialisation, then, people with impairments were more likely to be seen as burdens than contributing members of society, better catered for in institutions than the community – at least in principle, if not in practice.[15]

Although most clearly expressed and elaborated by historical materialists, this 'industrialisation thesis' about the conditions and forces responsible for the creation of modern Western 'disability' (as a distinct social category and experience) has passed uncritically into the work of many cultural disability studies scholars such as Rosemarie Garland Thomson.[16] The broad appeal of materialist inspired accounts of disability is easy to understand. By calling attention to the structural basis of disabled people's experiences, they usefully show how disability is constituted in concrete ways. Barriers to paid employment, for instance, undoubtedly affect disabled people's position in society, as does the accessibility of the built environment. The analytic value of the industrialisation thesis in all its various guises is that it suggests the importance of changing material conditions and how these have affected the lives of people with impairments through history. The problem, however, is that its use as an explanatory framework is undermined by its lack of an adequate empirical foundation. Ideas about the Industrial Revolution's impact on disabled people's place in the world of work are central to the industrialisation thesis, but there are very few historical studies exploring the topic. Those that exist, moreover, tend to examine Western nations other than Britain, such as the United States or Belgium.[17] However, while the employment prospects of impaired Britons during industrialisation have not yet received sustained investigation, their experiences are of immense significance to disability history and theory. As the world's first industrial nation, and an influential model for those that followed, Britain and its experience of industrialisation is crucial to our understanding of the origins and nature of disability in the West today.

The explanatory power of the industrialisation thesis in disability studies is also weakened, as Anne Borsay has noted, by its reliance on an over-simplified account of the Industrial Revolution that emphasises factory production at the expense of other sectors of the economy.[18] Industrialisation, however, was a multi-faceted and uneven process. Since the 1980s, economic historians have challenged the view that the Industrial Revolution marked a rapid and decisive shift to factory production and called into question the pace and impact of economic change.[19] Factories may have sprung up in increasing numbers, but they were generally confined to relatively discreet manufacturing districts. They were not a ubiquitous feature of industrialising society. More important and common aspects of industrialisation included the growth of urban settlements, the increasing use of waged labour, increased mobility, the emergence of a market economy and intensification in the exploitation of natural resources.[20] These broader dimensions of industrial change have rarely been studied from a disability perspective. This book therefore has a dual objective: to encourage historians of industrial society to incorporate experiences of disability into their analyses and to help disability scholars develop a more nuanced view of industrialisation by showing what can be gained when the focus of attention is shifted away from factories towards other important sites of industrial development – in our case, the mines and pit villages of industrialising Britain. Furthermore, people's relationship to work may be an important determinant of their social position and experiences as the industrialisation thesis maintains, but it is not the only one. In going beyond the workplace and looking at 'disabled' Britons' experiences in other areas of life during the Industrial Revolution, this book suggests how primarily economic meanings of disability could be mediated and challenged by, for example, disabled people's domestic, spiritual and social lives.

Indeed, those who witnessed the Industrial Revolution were far more concerned about the impact it had on the bodies of workers than what it meant for the employment prospects of 'disabled' people. Critics of mechanisation and reformers seeking to limit the employment of children in textile mills during the 1830s routinely pointed to the damaging effects of factory work on the health, posture and well-being of employees.[21] Some observers regarded conditions in collieries as even worse. A witness to the 1833 Factory Commission remarked that 'the hardest labour in the worst room in the worst-conducted factory is less hard, less cruel, and less demoralizing than the labour in the best of coalmines'.[22]

Bodily non-normativity *defined* workers in industrialising Britain. For instance, William Dodd, the self-styled 'factory cripple' who campaigned against exploitative conditions of work in the woollen mills of northern England, wrote in 1841 that various categories of industrial worker could be defined by their 'shape', from 'in-kneed cripples' to those whose legs 'curved both outwards, so that a person may run a wheel-barrow between them'. Both were the result of excessive standing in one position, or the 'over-exertion' that Dodd complained was endemic in textile mills.[23] Such claims were echoed in the critiques of industrial capitalism presented by Friedrich Engels and Karl Marx, who drew on government inquiries and factory inspector reports to show how new modes of production sacrificed the lives and limbs of workers.[24] As the century progressed, eugenicists also called attention to the diseases and deformities of workers to illustrate fears that living conditions in industrial cities would 'produce an inferior race of urban degenerates'.[25]

As Peter Kirby has cautioned, comments about the ubiquitous deformities or poor health of industrial workers were sometimes exaggerated and do not necessarily indicate the true scale of occupational disease and injury in industrialising Britain.[26] However, by highlighting the presence of *workers* with impairments they do call into question the claim that industrialisation made it hard for impaired people to take up productive economic roles. If the Industrial Revolution did indeed make 'disabled' people, it should also be remembered that disabled people also helped make the Industrial Revolution. Rather than passive bystanders or victims of industrialisation, therefore, disabled people were actually active agents of economic change, though this is rarely acknowledged.

Put simply, then, a new approach to disability and industrial society is needed – one that takes into account the multi-faceted nature of industrial change and explores the dynamics of inclusion and exclusion within particular cohorts of workers or occupations, and in different settings. Sofie De Veirman's recent work, using census and other records to explore the changing work experiences of deaf people in eighteenth- and nineteenth-century Belgium, offers one fruitful way forward.[27] By focusing on disability within a particular sector of the economy, rather than tracing the experiences of a single impairment group, we present another. Such an approach enables a more nuanced analysis of workers' experiences of impairment within the context of changing working practices, employer attitudes and industrial politics. It also takes into account the lives of disabled people beyond the workplace, to examine their familial, community and religious experiences. This moves us from a problematic general, all-encompassing theory of disabling industrialisation that privileges the world of work at the expense of other aspects of life, towards a history that makes room for the 'specific, local and personal'.[28]

Just as disability scholars and economic historians have failed to adequately recognise the productive contributions of disabled people in the past, so too have labour historians. Although occupational disease and injury have been major themes in labour history, labour historians hardly ever portray disabled people as workers. Instead, they seem to have assumed that, once injured, disabled workers simply left the workforce.[29] Not only does this obscure the historical meaning of work and impairment; it also reinforces inaccurate and harmful ideas about the productive capacities of disabled people. Sarah Rose and others have drawn attention to the significance of disability 'as a metaphor and as a key historical aspect of working-class life and communities'. For Rose, incorporating a 'disability perspective' into histories of labour and working-class life has the potential to transform our understanding of industrial relations, work, dependency and citizenship.[30] By critically engaging with the industrialisation thesis, this book similarly opens up new avenues of inquiry that have significant ramifications for fields beyond disability studies and the way disabled people are viewed today.

Coalmining in Britain, 1780–1880

Coalmining has been chosen as the focus of this study because it was central to the economic development of modern Britain and its workers were at high risk of occupational disease and injury. Not only did mining predate the coming of factories, it also helped shape their development in crucial ways. As E. A. Wrigley has argued, the increasing exploitation of fossil fuels from the late eighteenth century onwards removed significant barriers to the scale and location of industrial expansion.[31] Shifting from burning wood to coal removed competition for a natural resource upon which there were other demands that restricted its use. This enabled the widespread use of steam power and a move away from a reliance on water-driven technologies, freeing industry to expand into areas away from fast-flowing streams.[32] The transition to new sources of fuel began from the 1770s and between 1820 and 1870 coal was put to a wide range of industrial uses.[33] While coal was not the sole reason for the expansion of British manufacturing from the mid-eighteenth century, its near universal adoption in the nineteenth century certainly facilitated continuous growth in both production and population.[34] Coal powered the engines, mills and furnaces that drove the Industrial Revolution. It also fuelled the ships and locomotives that transported goods, materials, and people to far-off places, both within the British Isles and beyond. Consequently, British industry was able to exploit distant markets, as well as draw labour from further afield easier than ever before.

In addition, mining was a magnet to industry, enticing many industrialists whose businesses depended on huge quantities of coal to locate their enterprises in the coalfields.[35] The sinking of new pits led to the rapid expansion of new communities and an influx of people from far and wide. As an article in the *Penny Magazine* (1835) put it: '[i]f a new colliery is opened in a part of the country where such work had not previously existed, the colliery village springs up in necessary connexion with it.'[36] The speed and scale of population growth and urbanisation in mining areas is indicated by the example of the Easington district in north-east England. In November 1840, a government official reported that the population there had more than doubled since the mid-1830s due to the opening of new collieries and had necessitated the formation of a new town, Seaham Harbour.[37] The expansion of coalmining was a catalyst for demographic change and migration that were the hallmarks of industrial Britain. Rising demand for coal also stimulated technological innovations in coalmining itself and the growth of extensive transport networks to service mining, as coal owners sought to reduce their costs and supply far away markets.[38]

Coalmining's importance and dramatic expansion in the nineteenth century is powerfully illustrated by output figures for the period. Although national output statistics are approximate before the 1872 Coal Mines Act mandated better collection of production data, annual output in 1700 was estimated to have been just under three million tons, rising to roughly five million by 1750 and over thirty million by 1830.[39] By the 1870s output had reached 128 million tons and would rise even more dramatically to a peak of a little over 228 million by the outbreak of the First World War. Employment in the sector at this time grew at a similar rate. In 1800 an estimated 40,000 persons worked in the coal industry. By 1880, the figure stood at around 485,000.[40]

However, the expansion of the coal industry was uneven and marked by distinctive regional differences. As Figure 1 indicates, the main centres of coalmining in nineteenth-century Britain were north-east England, central

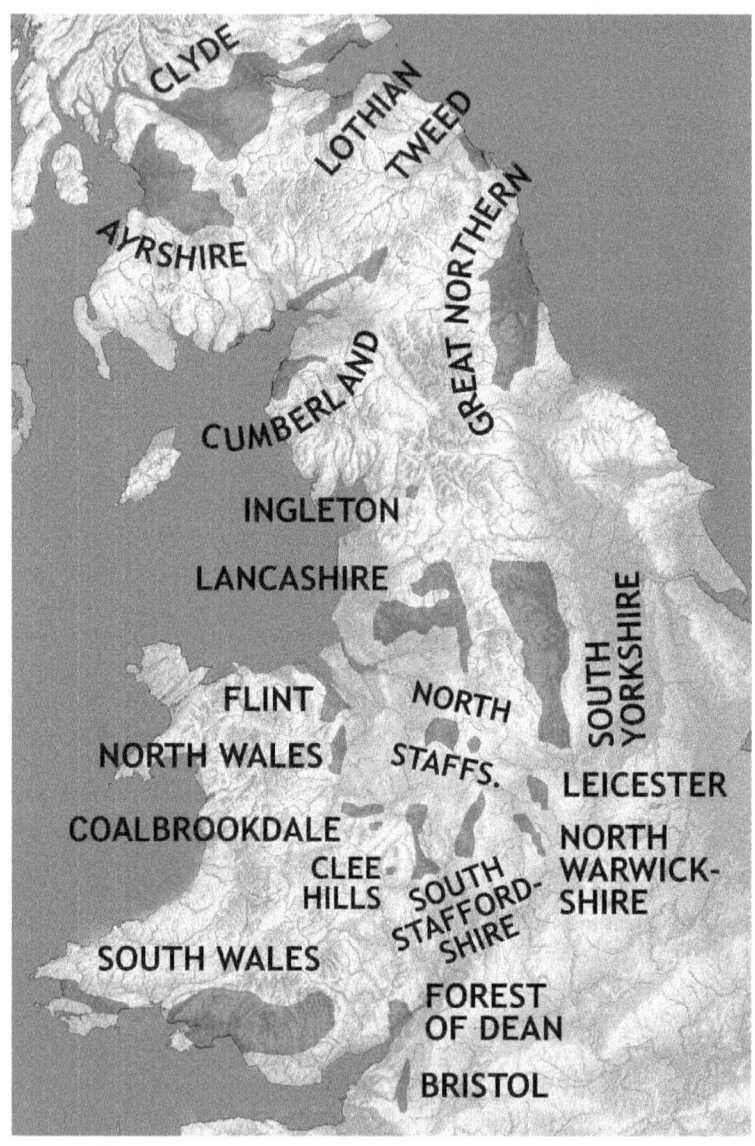

Figure 1 British coalfields in the nineteenth century, adapted from R. A. S. Redmayne, 'The Coal-Mining Industry of the United Kingdom', *The Engineering Magazine*, xxvi (1904). Credit: Notuncurious/Wikimedia Commons/CC-BY-SA-3.0.

Scotland, Lancashire, Yorkshire, the English Midlands and south Wales. This book pays special attention to those in central Scotland, north-east England, and south Wales, as these were three of the most significant coal producing regions, which reflect the varying pace of development of the industry during this period and the related differences in work cultures and conditions. Of these, the Great Northern Coalfield of north-east England was the oldest and largest, already employing 13,500 workers at the start of the nineteenth century. During the period 1780–1880, it remained the most important in Britain, both in terms of production and the numbers employed. However, north-east England's overwhelming dominance in mining, was gradually eroded during the course of the nineteenth century, due mainly to the dramatic rise of the south Wales coal industry from the 1840s. South Wales was the fastest growing coalfield during this period, producing about 20 per cent of the nation's coal by 1913. Scottish coalmining also experienced changing fortunes over the nineteenth century. In 1800 it accounted for around 20 per cent of the estimated total British coal production. Eighty years later, although its output had risen, its overall share of British output had fallen to 12 per cent, thanks to rapid expansion of mining in other regions.[41]

There were furthermore significant geological variations between (and within) coalfields that affected the rate, scale and methods of extraction, which in turn influenced the social organisation of mining communities.[42] Coal was a heterogeneous material. There were differences in the types and qualities of coal and the depths of mines needed to extract it. Bituminous coal was widely used for domestic heating and smelting metals. Usually found fairly close to the surface, it was relatively easy to mine and could be worked using comparatively cheap open-cast or drift mining techniques. Steam coal, in contrast, was further from the surface and more costly to mine.[43] As we shall see, geological differences between and within coalfields affected the risks coalminers faced since these influenced factors such as a mine's susceptibility to roof falls or the volume of dust inhaled by miners. Recognising the economic and cultural variations in Britain's coal industry during the Industrial Revolution illuminates the uneven nature of industrialisation, but it also compels us to see the diversity in 'disabled' people's experiences. As we shall see, the differences between and within the three coalfields chosen in this study were reflected in distinctive working practices, industrial relations and welfare provision for sick and injured miners that shaped the lives of disabled people in mining communities in multiple ways. By bringing in perspectives from different parts of Britain, this book further resists the homogenising impulse found in previous accounts of disability and the Industrial Revolution.

As mining expanded, British policymakers and the public became more aware of the shocking toll coal production took on the bodies of mineworkers. The number and scale of accidents in an industry regarded as vital to the economy, led to increasing regulation of the industry through official inspection.[44] At the same time, the question of how to provide for the many men, women and children badly hurt or 'worn out' in the service of mining inspired new welfare and medical responses. These initiatives helped make disability visible to British society and contributed to the idea, still popular today, that physical impairment is above all else a 'problem' that needs solving. Coalmining not only powered the Industrial Revolution, then, it also shaped emerging understandings and experiences of disability in nineteenth-century Britain that linger on, affecting the lives of disabled people in the present. Consequently, a disability history of British coalmining like this is long overdue.

Approach and methodology

While disability remains a neglected topic in histories of industrialisation, greater attention has been given to occupational diseases, working-class health and the regulation of nineteenth-century workplaces. Indeed, mining has occupied an important place in this scholarship. In his path-breaking book on *The Diseases of Miners* (1943), for example, George Rosen traced the evolution of medical thinking about the occupational health problems of mineworkers, particularly lung diseases.[45] Recent studies have added to Rosen's discussion by demonstrating that trade unions as well as doctors have been pivotal in producing medical knowledge.[46] Historians of occupational medicine have also highlighted industrial accidents and illnesses as evidence of the hardships faced by workers in the past, and the example of sick or injured workers has been used to explore working-class access to medical services via the Poor Law, employer paternalism, or voluntary benefit schemes.[47] For Paul Weindling, occupational health history provides a 'sensitive index of social conditions in industrial societies', and 'shows the power of the labour process to structure the everyday reality outside the workplace'.[48] Responses to the medical problems caused by the rapid expansion of urban industrial communities have been documented through studies of hospitals and other medical services.[49] Historians have also examined the nineteenth-century state's growing interest in regulating workplaces as part of a broader campaign for public health, linked to middle class reformism.[50] This has stimulated valuable research on the evolution of health and safety policy through nineteenth-century regulation of mining, factories and other 'dangerous trades', and on changing attitudes towards risk and accident prevention that led to new laws on workmen's compensation.[51]

The best of this work recognises that the history of workplace accidents and diseases is not simply a matter of medical diagnosis or public policy, but also about individual experiences. For example, in *Caught in the Machinery* (2008), Jamie Bronstein argues that the meanings of injury in industrialising Britain were formed in the 'multiple contexts in which it was experienced', and that the personal consequences of workplace accidents were contingent on many factors, including the nature of the injury, the skill of the worker involved and the attitudes of employers. She also examines the cultural representation of accidents – in mining as well as other sectors – and charts the relationship between work, injury and masculine identity.[52] However, her principal focus is on the evolution of Anglo-American employer liability law, which precludes further analysis of the ability of injured miners and other workers to return to employment, of how disability was experienced within working-class communities, or of the role played by injured or impaired workers themselves in industrial politics.

In this book we take a disability history approach to workplace accidents and ill health. This differs from conventional occupational health histories in that it seeks a more holistic approach to the experiences of injured and diseased workers – one that goes beyond the worlds of work and medicine to include consideration of other important aspects of disabled people's lives. Thus, the home, the family, the church, the courtroom and even the marital bed are equally important sites for investigation by disability historians as the workplace, hospital or poorhouse. Viewed from such a multi-dimensional perspective, 'disabled' people are seen as much more than mere patients, dependents or incapacitated workers. Instead, they become recognisable as the parents, spouses, brawlers, plaintiffs and respected members of their communities they also were. Highlighting and examining the diverse range of roles occupied by disabled people in the past enables their historical agency to be brought more fully into the spotlight, revealing the ways in which they have actively shaped their own lives and those of the people around them. Just as significantly, it also suggests how the multiple identities assumed by disabled people – as parents, workers, 'club' members and so on – affected their experiences of disability and its impact on their lives. Furthermore, although they were indeed the recipients of care, treatment or support, this was rarely, if ever, a top-down process, with health and welfare professionals wielding unrestrained power over disabled people. Instead, it was more often a process of negotiation marked variously by resistance, acquiescence and compromise on the part of both parties. Consequently, such circumstances enabled many disabled people to help fashion the treatment they received. Considering the effects of impairment from multiple perspectives, then, deepens our understanding of the contingent nature of disability and how it can mean very different things in different contexts and is always open to contest.[53]

In the pages that follow we focus primarily on physically disabling and chronic conditions that had an occupational basis in coalmining, such as amputation, mobility impairment, visual impairment and chronic illness, such as respiratory disease, which caused progressive 'debility'. We also examine the range of physical deformities and bodily markings associated with mining, from spinal curvature to scars and unusual facial features. Although our emphasis is on physical rather than mental impairment, we acknowledge that psychological illness did sometimes interact with physical conditions and that trauma and disease affected patients both physically and emotionally. Rather than engaging in retrospective diagnosis, the book explains how categories of physical disability were created and evolved. In this study the descriptive term 'disabled' is used to refer to people identified in the sources by various keywords of impairment, such as 'maimed', 'worn out', 'lame' or 'cripple', who potentially may have faced restrictions on their ability to perform everyday activities through injury, disease, congenital malformation, ageing or chronic illness, or whose appearance made them liable to be characterised by contemporary cultural ideas associated with non-normative bodies. Nevertheless, questions remain about whether the subjects of this book would necessarily have identified themselves as 'disabled'. Modern definitions of 'disability' and 'disabled' are the result of a process of historical development and do not map easily onto understandings of impairment in the past.[54]

While the terms 'disabled' and 'disability' were rarely used before the twentieth century to denote a group of people defined by their physical difference or as a 'master trope of human disqualification', the period covered by this book has been identified as marking significant changes that led to the development of modern categories.[55] Deborah Stone has argued that the growing use of the 'defective' category to categorise people with a range of physical, sensory and intellectual impairments from 'able-bodied' paupers within nineteenth-century English Poor Law administration, shows that people with physical difference were increasingly banded together as a 'problem' requiring specific responses.[56] Lennard Davis has also highlighted the significance of the era to the modern disability category, arguing that its roots can be traced to statisticians' attempts in the early nineteenth century to measure and define 'normal' (and by implication 'abnormal') human characteristics, which subsequently fed into Social Darwinism and eugenic ideas about improving the human species later in the century.[57] Despite such efforts to categorise human difference more rigorously, however, the term 'disabled' was often used in flexible ways during the formative phase of the Industrial Revolution. As

Henry de la Beche's use of the term in the example above indicates, 'disabled' sometimes referred to those whose ability to earn their living, and support themselves and their families, was compromised. At the same time, 'disabled' could also be used to describe someone who was unable to work at his or her usual occupation, rather than a person completely incapacitated from any kind of work whatsoever. Furthermore, whereas the modern term 'disabled' relates to a permanent or long-term condition, nineteenth-century Britons also used it to refer to temporary states.[58]

To understand 'disabled' people's experiences during this period of industrial expansion, we need to recognise the flexibility and subtlety of contemporary languages of bodily difference. The binary positions of 'ability' and 'disability' fail to capture the full range of people's experiences in the past.[59] In industrial societies where classes of workers were frequently defined by their distinctive physical peculiarities, and where accidents were common, it seems likely that many people occupied a liminal space between unimpaired physical wholeness and 'total' disablement. Take, for example, the person described by Dr James Mitchell in the 1842 Children's Employment Commission report, seriously injured, but eventually able to return to work 'a little distorted', mentioned at the start of this chapter. While he experienced lasting damage to his body, he did not fit a model of disability defined by complete physical incapacity. Chris Mounsey has argued that historical studies of disability need a new method for understanding bodily 'variability'. The uniqueness of a body, he argues is a result of its context, comprising three inter-related elements: a person's unique physical *capacity*; his or her *capability* to come to terms with his or her difference; and how these were affected and shaped by *encounters* with others.[60] It follows, therefore, that diseased or injured coal workers' experiences were affected by the nature of their impairment and its impact on their functional abilities, by their ability to accommodate themselves to this difference and by the approach of their employers, families and communities to this difference. The latter was manifested (among other ways) by colliery managers' willingness to accept this person as an employee or find alternative work for them, by the injured or impaired person's status within her or his home or community, and by the broader set of cultural values that shaped responses to those whose bodies did not necessarily conform to the qualities deemed desirable or acceptable within these communities. Although not all these features or interactions are easily visible in the historical record, the flexibility of this method allows us to put emphasis on the uniqueness of individual experiences and avoids the homogenising and inaccurate tendency to view 'disabled' people as a 'group' in industrialising Britain. However, at the same time, it is also important to acknowledge the commonality of experience that has existed between disabled people in the past, despite their different personal circumstances. For example, as this book shows, injured mineworkers may have had very different impairments, capacities, skills and employers, but they still often faced the same moralistic attitudes in their dealings with friendly societies or Poor Law officials and these could affect their lives in very similar ways. In the pages that follow, then, we recognise *both* the common and particular features of disabled people's lives and explore how these interacted with each other to shape their experiences.

Given the fluid and variable nature of 'disability', it's difficult to quantify the number of 'disabled' people in nineteenth-century Britain. In coal, as in other industries, these difficulties are compounded by a lack of reliable statistics for accidents and occupational diseases for the period. As Chapter 1 shows, although the growth of government regulation of mining after 1850 compelled mine owners and managers to report 'serious' accidents, fatal accidents were generally better reported than non-fatal ones. There was no official mechanism for measuring the incidence of occupational diseases, or psychological trauma either. As late as 1878, member of parliament (MP) Joseph Cowen of Newcastle noted that while the 'Home Secretary received a list of the persons who were absolutely killed, [in coal mines] ... he received no return as to the number of men who had their backs injured, their ribs squeezed in, or their legs broken'.[61] Mine inspectors, moreover, did not routinely collect data for time taken off work because of occupational disease and injury until the end of our period – when growing concerns among economists, medical professionals and policymakers about the effects of work-related incapacity on productivity stimulated more comprehensive documentation of time lost to sickness or injury.[62] Consequently, this book takes a cultural approach to disability in industrial society that focuses on the meanings of impairment rather than quantification. Moving away from the diagnostic approach in traditional medical histories, it explores how impaired workers saw themselves, asking what scars, missing limbs, sight loss, breathing difficulties and other injuries or chronic conditions meant for those who experienced them, and how they were perceived by others.

Such an approach demands close reading of a wide variety of texts and this book uses a rich and innovative mix of sources to examine disability and its consequences in three targeted coalfields. Parliamentary debates, legislation and official reports provide a wealth of information – often neglected by historians of disability – about the working conditions in mines, the health of workers and exposure to accidents. The 1842 Children's Employment Commission Report, for example, includes numerous descriptions and first-hand testimonies of colliery employees of all ages with a variety of physical impairments and medical conditions, as well as the testimonies of medical men and colliery managers. Providing the most complete picture of conditions in coal mines in the middle of our period, it is used extensively in this book. Although some of the claims made by expert witnesses to official inquiries were dubious, they nonetheless help us to map out the conflicting claims that shaped public and political understandings of coal workers' health and well-being.[63] Press reports and autobiographies also furnish evidence for the home life, community relations and political activities of mineworkers, including those with impairments. Methodist periodicals provide rich, if idealised, biographical accounts of miners and other workers that reveal the religious meanings attached to impairment, while newspapers of the labour movement indicate the political import of occupational injury and how it shaped miners' attitudes and approaches to industrial relations and state regulation.

The diseases of miners are explored via medical journals, treatises and hospital records. Although they foreground professional perspectives, they also reveal ways in which patients took their own decisions about their treatment. The complex responses to the welfare needs of those injured or disabled in the coal industry are documented in Poor Law papers and in friendly society records, which provide evidence of the ways in which miners took responsibility for planning for their own ill health or disability. Documents recording distinctive responses to disability in the coalfields, such as the Northumberland and Durham Miners' Permanent Relief Fund established in 1862 are examined in order to shed light on miners' self-conscious distancing from state-funded welfare under the Poor Law. Taken together, these sources open up new angles of vision on how ideas about disability were created and mediated in various social, textual and administrative settings. Reading institutional and administrative records 'against the grain' and alongside a variety of sources produced in other contexts, allows us to tease out and analyse the distinctive responses to disability among miners, their families and communities.

Chapter outline

This book comprises five thematic chapters. These examine the economic, medical and welfare responses to disease, injury and impairment among coal workers, and discuss experiences of disability within the context of social relations and the industrial politics of coalfield communities. Chapter 1 provides the context for those that follow by providing an overview of the conditions of work in British coalmining between 1780 and 1880. It pays close attention to the nature and variety of work at collieries and in coalfield settlements in order to better understand how economic conditions shaped the in/ability of disabled workers to remain productive. Chapter 2 turns its attention to the principal causes of disablement in the nineteenth-century coal industry and the medical responses to them. Often admired for their physical prowess, miners appeared to embody an ideal of working-class able-bodiedness. But increasingly, from the end of the eighteenth century, the risks to health posed by coalmining began to attract medical attention. Mineworkers, moreover, were one of the first sections of the industrial working population to become accustomed to the services of doctors and surgeons via colliery sick clubs, and their interaction with medical professionals helped shaped subsequent working-class experiences of medical care. Chapter 3 extends the discussion of responses to disability by examining the welfare provisions for miners with long-term restrictive health conditions. It examines how miners and their families negotiated a 'mixed economy' of welfare, comprising family and community support, the Poor Law, and voluntary self-help as well as employer paternalism.

Chapter 4 shifts attention away from medicine and welfare towards the ways in which disability affected social relations within coalfield communities. It explores how disability shaped the identities of men and women, focusing in particular on the community, family and religious life. Chapter 5 explores the place of disability in industrial politics and how fluctuating industrial relations affected the experiences of disabled people in the coalfields. It examines how labour leaders and their allies used the rhetorical figure of the disabled miner to advance a range of causes from better provision for injured mineworkers to improved working conditions more generally. While such representations often emphasised the suffering associated with disability, the chapter shows that impaired miners were rarely passive victims. On the contrary, it demonstrates that many took an active role in industrial politics. The chapter concludes by considering workers' successful fight for better compensation laws and the impact this had on the employment prospects of disabled miners.

Notes

1 Edward Kentish, *An Essay on Burns, In Two Parts: Principally on those Which Happen to Workmen in Mines from Explosions of Carburetted Hydrogen Gas* (London: Longman et al. 1817), Part 1, 129, 133.
2 Helen and Baron Duckham, *Great Pit Disasters. Great Britain, 1700 to the Present Day* (Newton Abbott: David and Charles, 1973).
3 John Benson, 'Non-Fatal Coalmining Accidents', *Bulletin of the Society for the Study of Labour History*, 32 (1976), 20–2.

4 PP 1842 (381), *Appendix to the First Report of the Commissioners. Mines. Part 1. Reports and Evidence from Sub-commissioners*, 138–9, 140. Our emphasis.
5 PP 1849 (613), *Report from the Select Committee of the House of Lords Appointed to Inquire into the Best Means of Preventing the Occurrence of Dangerous Accidents in Coal Mines*, 16.
6 For current approaches, see Emma Griffin, *A Short History of the British Industrial Revolution* (Basingstoke: Palgrave Macmillan, 2006); Steven King and Geoffrey Timmins, *Making Sense of the Industrial Revolution: English Economy and Society 1700–1850* (Manchester: Manchester University Press, 2001).
7 M. W. Flinn, *The History of the British Coal Industry*, vol. 2: *1700–1830: The Industrial Revolution* (Oxford: Clarendon Press, 1984); Roy Church, *The History of the British Coal Industry*, vol. 3: *1830–1913: Victorian Pre-Eminence* (Oxford: Clarendon Press, 1986).
8 P. E. H. Hair, 'Mortality from Violence in British Coal Mines, 1800–50', *Economic History Review*, 21:3 (1968), 545–61; John Benson, *British Coalminers in the Nineteenth Century: A Social History* (Dublin: Gill and Macmillan, 1980), 40.
9 Ibid., 43.
10 Ibid., 44. Cf. Rosemarie Garland Thomson (ed.), *Freakery: Cultural Spectacles of the Extraordinary Body* (New York: New York University Press, 1996); Nadja Durbach, *Spectacle of Deformity: Freak Show and Modern British Culture* (Berkeley: University of California Press, 2008).
11 P. W. J. Bartrip and S. B. Burman, *The Wounded Soldiers of Industry: Industrial Compensation Policy 1833–1897* (Oxford: Clarendon Press, 1983), ch. 5.
12 Deborah Stone, *The Disabled State* (Basingstoke: Macmillan, 1984); Lennard J. Davis, *Enforcing Normalcy: Disability, Deafness and the Body* (London: Verso, 1995).
13 Vic Finkelstein, 'Disability and the Helper/Helped Relationship. An Historical View', in Ann Brechin et al. (eds), *Handicap in a Social World* (Sevenoaks: Hodder & Stoughton in association with the Open University Press, 1981), 65–78; Michael Oliver and Colin Barnes, *The New Politics of Disablement* (Basingstoke: Palgrave Macmillan, 2012), 52–73; Brendan Gleeson, *Geographies of Disability* (London: Routledge, 1999); Nirmala Erevelles, 'In Search of the Disabled Subject', in James C. Wilson and Cynthia Lewiecki-Wilson (eds), *Embodied Rhetorics: Disability in Language and Culture* (Carbondale and Edwardsville: Southern Illinois Press, 2001), 100–6; Penny L. Richards, 'Industrialization', in Susan Burch (ed.), *Encyclopedia of American Disability History*, 3 vols (New York: Facts on File, 2009), ii, 482–3; Lennard J. Davis, *Enforcing Normalcy: Disability, Deafness, and the Body* (London: Verso, 1995), 86–90.
14 Gleeson, *Geographies of Disability*, 80–98.
15 Oliver and Barnes, *New Politics of Disablement*, 52–73.
16 Rosemarie Garland Thomson, 'Introduction: From Wonder to Error – a Genealogy of Freak Discourse in Modernity', in Garland Thomson (ed.), *Freakery*, 11–12.
17 For example, Daniel Blackie, 'Disabled Revolutionary War Veterans and

the Construction of Disability in the Early United States, c. 1776–1840' (unpublished PhD thesis, University of Helsinki, 2010), http://urn.fi/URN:ISBN:978-952-10-6343-5, 109–40; Sarah F. Rose, *No Right To Be Idle: The Invention Of Disability, 1840s–1930s* (Chapel Hill: University of North Carolina Press, 2017); Halle Gayle Lewis, '"Cripples Are Not the Dependents One Is Led to Think": Work and Disability in Industrializing Cleveland, 1861–1916' (unpublished PhD thesis, State University of New York at Binghamton 2004); Sofie De Veirman, 'Deaf and Disabled? (Un)Employment of Deaf People in Belgium: A Comparison of Eighteenth-Century and Nineteenth-Century Cohorts', *Disability and Society*, 30:3 (2015), 460–74.

18 Anne Borsay, *Disability and Social Policy in Britain since 1750: A History of Exclusion* (Basingstoke: Palgrave Macmillan, 2005), 10–15; Gleeson, *Geographies of Disability*, 106–7; Oliver and Barnes, *New Politics*, 54–7, 60–1.

19 Griffin, *Short History*, 6; Maxine Berg and Pat Hudson, 'Rehabilitating the Industrial Revolution', *Economic History Review*, 45:1 (1992), 24–50.

20 John Williams, *Was Wales Industrialised? Essays in Modern Welsh History* (Llandysul, Dyfed: Gomer Press, 1995), 14–36; Griffin, *Short History*, chs 3 and 4.

21 Peter Kirby, *Child Workers and Industrial Health in Britain, 1780–1850* (Woodbridge: Boydell Press, 2013); Pamela Sharpe, 'Explaining the Short Stature of the Poor: Chronic Childhood Disease and Growth in Nineteenth-Century England', *The Economic History Review*, 65 (2012), 1475–94.

22 PP 1833 (450), *Factories Inquiry Commission. First Report of His Majesty's Commissioners Appointed to Collect Information in the Manufacturing Districts, as to the Employment of Children in Factories*, 82.

23 William Dodd, *A Narrative of the Experiences and Sufferings of William Dodd, A Factory Cripple*, 2nd edition (London: L. and G. Seeley, 1841), 29, 30.

24 Steffan Bengtsson, 'Out of the Frame? Disability and the Body in the Writings of Karl Marx', *Scandinavian Journal of Disability Research*, 19:2 (2016), 151–60.

25 Paul Weindling, 'Linking Self Help and Medical Science: the Social History of Occupational Health', in Paul Weindling (ed.), *The Social History of Occupational Health* (London: Croom Helm, 1985), 7.

26 Kirby, *Child Workers*, 5, 12. See also P. W. J. Bartrip and S. B. Burman, *The Wounded Soldiers of Industry: Industrial Compensation Policy 1833–1897* (Oxford: Clarendon Press, 1983), 10–13.

27 De Veirman, 'Deaf and Disabled?'.

28 Chris Mounsey, 'Introduction: Variability: Beyond Sameness and Difference', in Chris Mounsey (ed.), *The Idea of Disability in the Eighteenth Century* (Lewisburg, PA: Bucknell University Press, 2014), 5.

29 Sarah F. Rose, '"Crippled" Hands: Disability in Labor and Working-Class History', *Labor*, 2:1 (2005), 27–54.

30 Ibid., 29, 47. See also Edward Slavishak, *Bodies of Work: Civic Display and Labor in Industrial Pittsburgh* (Durham, NC and London: Duke University Press, 2008).

31 E. A. Wrigley, *Continuity, Chance and Change: the Character of the Industrial Revolution in England* (Cambridge: Cambridge University Press, 1990).
32 Griffin, *Short History*, 107, 114.
33 Ibid., 125.
34 Ibid., 114.
35 Gerard Turnbull, 'Canals, Coal and Regional Growth during the Industrial Revolution', *Economic History Review*, 40 (1987), 537–60.
36 'The Collieries – No. 1', *Monthly Supplement of the Penny Magazine of the Society for the Diffusion of Useful Knowledge*, 28 February –1 March 1835, 123.
37 TNA, MH12/3052, Easington Union, Poor Law Correspondence, 1834–45, Letter from District Poor Law Commission Representative to the Poor Law Committee, 13 November 1840.
38 Flinn, *British Coal Industry*; J. H. Morris and L. J. Williams, *The South Wales Coal Industry, 1841–1875* (Cardiff: University of Wales Press, 1958).
39 Church, *British Coal Industry*, 2; Flinn, *British Coal Industry*, 26, table 1.2.
40 Benson, *British Coalminers*, 7.
41 Ibid., 9–27, 216–17; Flinn, *British Coal Industry*, 5–23.
42 Stefan Berger, 'Introduction' in Stefan Berger, Andy Croll and Norman Laporte (eds), *Towards a Comparative History of Coalfield Societies* (Aldershot: Ashgate, 2005), 2.
43 Benson, *British Coalminers*, 17.
44 Catherine Mills, *Regulating Health and Safety in the British Mining Industries, 1800–1914* (Farnham: Ashgate, 2010).
45 George Rosen, *The History of Miners' Diseases: A Medical and Social Interpretation*, with an Introduction by Henry E. Sigerist (New York: Schuman's, 1943).
46 For example, Arthur McIvor and Ronald Johnston, *Miners' Lung: A History of Dust Disease in British Coal Mining* (Aldershot: Ashgate, 2007).
47 For example, Leah Leneman, 'Lives and Limbs: Company Records as a Source for the History of Industrial Injuries', *Social History of Medicine*, 6:3 (1993), 405–27; David G. Green, *Working Class Patients and the Medical Establishment: Self-Help in Britain from the Mid-Nineteenth Century to 1948* (Aldershot: Gower, 1985); James C. Riley, *Sick Not Dead: The Health of British Workingmen during the Mortality Decline* (Baltimore, MD: Johns Hopkins University Press, 1997).
48 Weindling, 'Linking Self-Help and Medical Science', 2, 18.
49 For example, John V. Pickstone, *Medicine and Industrial Society: A History of Hospital Development in Manchester and Its Region 1752–1946* (Manchester: Manchester University Press, 1985).
50 Anthony S. Wohl, *Endangered Lives: Public Health in Victorian Britain* (London: J. M. Dent, 1983); Robert Gray, 'Medical Men, Industrial Labour and the State in Britain, 1830–50', *Social History*, 16:1 (1991), 19–43.
51 Mills, *Regulating Health and Safety*; Bartrip and Burman, *Wounded Soldiers of Industry*; P. W. J. Bartrip, *The Home Office and the Dangerous Trades: Regulating Occupational Disease in Victorian and Edwardian Britain* (Amsterdam: Rodopi, 2002).

52 Jamie L. Bronstein, *Caught in the Machinery: Workplace Accidents and Injured Workers in Nineteenth-Century Britain* (Stanford, CA: Stanford University Press, 2008), 4, 87 and ch. 3 *passim*. For a similar approach, see Slavishak, *Bodies of Work*.
53 Beth Linker, 'On the Borderland of Medical and Disability History: A Survey of the Fields', *Bulletin of the History of Medicine*, 87:4 (2013), 499–535; Susan Burch and Michael Rembis (eds), *Disability Histories* (Urbana: University of Illinois Press, 2014).
54 Roger Cooter, 'The Disabled Body', in Roger Cooter and John Pickstone (eds), *Companion to Medicine in the Twentieth Century* (London and New York: Routledge, 2003), 367–84.
55 David T. Mitchell and Sharon Snyder, *Narrative Prosthesis: Disability and the Dependencies of Discourse* (Ann Arbor: University of Michigan Press, 2000), 3.
56 Stone, *Disabled State*, 29–55.
57 Davis, *Enforcing Normalcy*, 23–49.
58 UK Government, 'Definition of Disability Under the Equality Act 2010', https://www.gov.uk/definition-of-disability-under-equality-act-2010, accessed 14 March 2017; cf. David M. Turner, *Disability in Eighteenth-Century England: Imagining Physical Impairment* (New York: Routledge, 2012), ch. 1. For the fluidity of nineteenth-century uses of 'disabled', see Chapter 3.
59 Catherine J. Kudlick, 'Comment: On the Borderlands of Medical and Disability History', *Bulletin of the History of Medicine*, 87:4 (2013), 551.
60 Mounsey, 'Introduction', 18.
61 *Hansard*, HC Deb 21 June 1878, vol. 241, cols 84–5.
62 Steve Sturdy, 'The Industrial Body', in Cooter and Pickstone (eds), *Companion to Medicine*, 220–1. PP 1881 [C.2903] *Mines. Reports of the Inspectors of Mines, to Her Majesty's Secretary of State, for the Year 1880*, 196–203.
63 Kirby, *Child Workers*, 5.

I

DISABILITY AND WORK IN THE COAL ECONOMY

Thomas Burt's early memories of mining were haunted by the sight of the mutilated bodies of his fellow workers. Remembering his work as a teenage pony putter in the 1850s, responsible for moving coal underground at Murton Colliery, County Durham, Burt recalled that 'everywhere, below ground and above, dangers stood thick'. Compounded by the 'rush and recklessness' of workers there, these dangers meant accidents were common. 'Never', he wrote in his autobiography published posthumously in 1924, 'had I seen so many crutches, so many empty jacket sleeves, so many wooden legs.' Returning to his native Northumberland after his Murton experiences to work at Cramlington Colliery, Burt was again struck by the high frequency of 'accidents to life and limb' and noticed people using 'crutches and wooden legs' among the local population. Although these were 'less conspicuous than at Murton', workers with impairments were a common sight at Cramlington. Burt recorded at least one by name, a mineworker called Bob Barrass who had 'unhappily lost an eye' but worked as a rolley-way man, maintaining the underground roads on which coal was transported. The dangers at Cramlington were so 'great' and the incidence of injury so high that Burt regarded it as remarkable that both he and his father 'were fortunate enough to leave the place unscathed and uninjured'.[1]

Written from the perspective of his later role as an MP and trade unionist, Burt's comments about the high number of injured workers he encountered in mining were intended to highlight the dangers faced by miners in pits where the safety of workers seemed to matter less to colliery owners than profit. Yet, beyond the political and rhetorical use to which he put them, Burt's observations also raise important questions about the supposed consequences of industrialisation for 'disabled' people. As Burt's recollections make clear, workers with impairments were not automatically forced from the mines during the Industrial Revolution but continued to work there in considerable numbers despite their injuries. If industrialisation was such a calamity for disabled people's working lives, as some disability scholars have argued, how were so many people with impairments able to work in a sector crucial to Britain's industrial economic development? And how far, if at all, did they, or others, actually regard themselves as 'disabled' people?

This chapter addresses these questions by examining the nature of mine work and the development of mining in the nineteenth century, paying special attention to the factors that enabled injured workers to participate in the working life of collieries and the extent to which they did so. To understand perceptions and experiences of disability during industrialisation it is necessary to examine the nature of 'industrial work' in all its forms. This chapter facilitates such an assessment by specifically exploring the ways in which economic factors, from working techniques and changing technologies to contracts and conditions of employment, affected the perception and experiences of injured and impaired mineworkers. It explores the extent of disablement in coalmining and analyses the working experiences of people with impairments within coal mines and their surrounding communities. In the process, it re-evaluates the relationship between 'disability' and work in industrialising Britain and suggests that popular ideas about the impact of the Industrial Revolution on disabled people's lives that emphasise their exclusion from work need re-thinking.[2]

The nature and conditions of mine work

Mineworkers' experiences in the coal industry were shaped by the differing economic trajectories and geologies of the specific coalfields in which they worked, as well as the significant cultural differences between them. The most glaring of these, particularly in the early nineteenth century, concerns the division of colliery labour and the employment of women and girls. According to the report of the commission set up to investigate children's employment in mines published in 1842, 'female Children of tender age and young and adult women are allowed to descend into the coal mines and regularly ... perform the same kinds of underground work, as boys and men' at some collieries in Yorkshire and Lancashire.[3] The practice of female work in south Wales' early nineteenth-century mines was 'not uncommon' either, but was most prevalent in Pembrokeshire, where women worked on the surface and underground usually operating a windlass by which loads of coal were drawn up steep passages.[4] Female mine work had been declining since the start of the nineteenth century and in some areas, such as north-east England, women were already excluded from working underground.[5] By the time of the Children's Employment Commission only an estimated 4 per cent of *all* British workers in coalmining were female.[6] Of these women and girls, however, a substantial proportion (2341, or about 40 per cent of the total) worked at Scottish mines, particularly in the eastern part of the coalfield.[7] Here, women and girls were heavily involved in underground work and were employed in hauling and bearing coal in significant numbers.[8] Women's work in early nineteenth-century Scottish coal mines was a legacy of the system of serfdom in Scottish coalmining, which had lasted from 1606 until 1799, where whole families had been bound to mine owners for life.[9]

The Mines and Collieries Act passed in August 1842 prohibited all females and boys under the age of ten from working underground.[10] Women continued to work in various capacities above ground at collieries, but their involvement in underground tasks ceased.[11] Although the numbers of women working at collieries at each census between 1841 and 1911 remained fairly consistent at between 4000 and 6000, the female proportion of the total workforce declined from around 3.5 per cent in 1841 (just before underground work became illegal) to less than 1.25 per cent after 1861.[12] Moreover, although the law prohibiting boys under ten from working below ground was sometimes flouted, the 1842 Act meant that the age profile of the underground workforce became more mature.[13] While underground mine work had arguably always been seen as a masculine domain – even in Scotland where one female miner reported that 'men about this place don't wish wives to work in mines but the masters seem to encourage it' – the gendering of coalmining as men's work was, after 1842, reinforced by government regulation.[14]

There were many different jobs at collieries, on the surface as well as below ground. An account of South Hetton Colliery near Durham published in the *Penny Magazine* in 1835 indicates the diversity of occupations at a single pit. At the top of the hierarchy were the colliery manager, viewer (an agent or surveyor appointed by the owner to run the colliery), first and second engineer and surgeon. Above ground worked thirteen joiners and sawyers, seven engine-wrights who made and repaired the pit's machinery, eight engine-men employed to keep the machinery working, nine firemen to attend the boilers, eighteen smiths to prepare the iron work required in the machinery and wagons, eight masons, six labourers, eleven cartmen, nine horsemen and a saddler. Other employees at the surface included banksmen (who emptied the tubs or corves of coal), boys employed to pick out stones and clean the coals, and railway attendants. Of the colliery's 526 employees, 210 worked above ground; the rest worked below at cutting or hewing coal, hauling it, or as underground foremen and support workers.[15]

By the end of the nineteenth century there were more than 200,000 men, women and children working at the pithead. This work, as John Benson has demonstrated, could be heavier, dirtier, more unpleasant and dangerous than many other jobs above ground in Victorian Britain. While rarely considered in eighteenth- and nineteenth-century discussions of occupational morbidity in the coal industry, surface workers faced a number of threats to their health and well-being, such as inhalation of coal dusts, the danger of various accidents such as being run over by wagons, and ruptures and strains from heavy lifting.[16] The variety of roles available at mines gave rise to an occupational hierarchy within the coal industry that was reflected in different levels of pay and status between jobs. The fundamental division in colliery work was between those who worked above ground and those who worked below. Working below tended to attract higher pay, was regarded as higher status and, as we shall see, carried considerable risk of accident.

Of the numerous non-supervisory jobs below ground, hewing, or coal-cutting, was at the top of the occupational hierarchy. It commanded the best wages and had the highest occupational status throughout Britain's coalfields. As former County Durham pitman George Parkinson recalled with great pride, hewing was 'the highest unofficial position attainable' underground.[17] In Scotland and north-east England, hewers regarded themselves as skilled craftsmen rather than common or unskilled workers and compared themselves to other skilled artisans such as stonemasons.[18] The status of the hewer was reinforced by the belief that, in Parkinson's words, he had the 'hardest' of mining jobs.[19] It was hard in two senses. First, it was physically arduous. As the *Colliery Guardian* observed,

> The hewer sits on a low stool (four inches in height) and grasping his pick with both hands, makes successive horizontal blows. To give the greatest effect to the stroke, his head is thrown back to one side, his left leg extended and his right bent, his right elbow resting on the right thigh, enables the leg to augment the force of the arm.[20]

Physical exertion was common to all hewing, but the degree of physical toil was affected by local geological conditions. The differing softness or thickness of the seam was also important in determining the physical demands on the bodies of miners. 'The differences of thickness sometimes admit the erect posture,' noted the *Penny Magazine*, but often men were obliged to 'sit, recline, or bend the body to an extreme degree.' The exploitation of deeper seams increased the temperature in which hewers worked, a situation exacerbated by having to work in very narrow, confined spaces in some collieries.[21] As we shall see in the following chapter, the health of coalminers was frequently judged by their posture and the deleterious effects of working in contorted positions for long periods of time.

Second, coal-cutting not only necessitated physical strength and endurance, it also entailed technical skill on the part of the hewer and 'considerable dexterity and experience'.[22] Generally, large coals fetched a better price at market and were the most prized, both by colliers and mine owners. To get these, miners had to work the coalface with great care, so as not to break slabs of coal into less valuable small coals.[23] Being a successful hewer depended on the ability to use a variety of tools and interpret the geological and other conditions in which a miner worked.[24] Hewing may have been physically strenuous, but it was not just a matter of brute force that simply favoured the most able-bodied above all others. It was from a combination of the *strength* and *skill* required by their job that hewers derived their strong occupational identity and pride.[25]

Nevertheless, the earnings of mineworkers depended on their capacity for hard physical toil. Colliers throughout Britain's coalfields were generally paid at piecework rates, which based pay on the amount they produced rather than the hours they worked. In north-east England, where miners were primarily employed to cut coal and nothing else, the relationship between coal produced and earnings was fairly simple. In south Wales, the terms under which colliers worked meant their earning potential was not so straightforward. While they too were paid for how much coal they produced, they also received remuneration for much of the non-cutting support work they performed. Colliers in south Wales were responsible for propping and maintaining the area around their individual workplaces, whereas in north-east England hewers usually left this to ancillary workers. In addition to this, Welsh colliers loaded their produce into coal trams for transportation to the surface.[26] The pay for these tasks tended to be determined on a piece rate basis as well. The problem with this was that ancillary work usually paid much less than coal-cutting. Consequently, Welsh colliers were under financial temptation to minimise the time they spent on 'offhand' or 'deadwork' (such as propping) and concentrate their efforts on hewing, which was better rewarded. Although less profitable, however, ancillary work was vital to safety. Without adequate propping, for instance, underground work spaces were more susceptible to roof falls. It is possible, therefore, that the organisation of work in the south Wales coal industry, combined with its remuneration practices, operated in a way that increased the risks to which Welsh miners were exposed and contributed to its notoriety as a particularly dangerous coalfield.[27]

Hewers made up a significant proportion of the underground work force at British collieries in the nineteenth century. Benson estimates that around half of those employed below ground were involved in cutting coal.[28] The other half worked in support roles. Ancillary tasks included hauling and bearing coal, and trapping, or operating the underground doors important for mine ventilation. Although mines required some skilled support workers, such as those who maintained the tracked roadways on which coal trucks were transported, much underground support work was heavy and unskilled in nature.[29] Both before and after the Mines and Collieries Act of 1842, hauling and trapping tended to be undertaken by the youngest sections of the underground workforce, although after the 1842 Act haulage work was increasingly done (where passages were wide enough) by pit ponies rather than human bearers. In north-east England, the hierarchy of underground work mapped onto the age of workers. As a witness to the Parliamentary Select Committee on Accidents in Coal Mines of 1835 explained, the youngest workers began as trap door keepers (12–14 years), stronger boys moved on to work as rolley drivers (responsible for moving wheeled sledges of coal known locally as 'rolleys'), before moving on to putting (hauling coal through underground passageways by any means).[30] The transition from putter to hewer was an important stage in a miner's life cycle, marking the transition from youth to adulthood. When he began hewing at Seaton Delaval Colliery from the age of eighteen, Thomas Burt remarked that he 'now ceased to be a boy and henceforth was a man'.[31]

In south Wales, the progression to coal-cutting and the distinction between face and 'offhand' work was less clear. Despite this, there was still some correlation between age and position. Young Welsh mineworkers served a more traditional form of apprenticeship than their counterparts in north-east England, learning their trade alongside a more experienced collier as his 'lad'. Although he may have performed some hewing as well as support tasks, a collier's lad generally spent more time engaged in ancillary work than cutting. It was not until their early twenties that Welsh 'lads' usually became full colliers in their own right.[32] The south Wales miner Edmund Stonelake, who started working underground in the 1880s, described how, in the absence of his deceased father, his mother had searched for a collier with whom he could be placed who could 'look after my safety, discipline me, and shelter me from bad company'. After initially being placed with an 'illiterate drunken Cornishman', he was transferred to a 'proper gentleman collier, a tall smart elderly man who kept himself aloof from other colliers, in the work and out of it'.[33] Although the ages at which youths became hewers varied, adult men normally undertook the hard, skilled and prestigious work of coal-cutting.

Just as there was much diversity in the type, nature and difficulty of jobs available at collieries, so too was there great diversity in the terms of employment governing mineworkers. Pay and contracts differed widely not only from coalfield to coalfield, but also within them. Miners had a deserved reputation as well-paid workers. Commenting on their situation in 1798, Thomas Gisborne noted that an average miner was consistently able to earn more than an agricultural labourer.[34] Hewers were the best paid of mineworkers, but throughout the period all underground workers were relatively well paid compared to other manual workers in agriculture, building and railways as well as non-agricultural labourers.[35] This was due to the greater risk of death or disablement involved in underground work and, to an extent, the related difficulties some employers faced in attracting people to work there.[36] In the first part of the nineteenth century, political economists argued that wages ought to reflect and compensate for any dangers that employees might face in the workplace.[37] Coal owners were especially keen to make a connection between the risks of mining and the relatively high wages they paid to their workers and, as we shall see in subsequent chapters, often supplemented these with paternalist benefits such as workplace medical schemes.[38]

Nevertheless, in spite of the theoretical link between rates of pay and exposure to the threat of death or disablement, the structure of miners' earnings was shaped by a variety of factors that went beyond the question of risk itself. All mineworkers were potentially subject to varying economic fortunes due to the vagaries of the market, whereas the temporary closure of pits due to flooding or explosions, could put many out of work. Individual earnings (as opposed to wages) varied by the number of shifts someone was able to undertake.[39] Hewers' earnings were determined by the geological structure of seams and local custom. Those working at cutting coal in north-east England tended to be paid better than those doing the same work elsewhere.[40] Moreover, there were differences within regions as well as between them. Miners working at collieries attached to iron works generally received less pay than their counterparts working in the sale coal sector, but were generally less susceptible to the vicissitudes of market fluctuations thanks to the more constant demands of the blast furnaces.[41]

Pay was not the only term of employment for mineworkers that varied widely from place to place. There was much diversity in the nature of contracts generally. Their length, for instance, could range from one month or less to several years. The three coalfields had quite different traditions in this area. Until the abolition of the bond system in 1872, mineworkers in north-east England entered into a legally binding agreement with their employers, usually on an annual basis, though there were some years when monthly bonds were used instead, particularly after the unsuccessful miners' strike of 1844. In principle, this system of hiring gave pitmen some security of employment, providing they kept to the terms of their bond, which were often onerous. For example, workers faced imprisonment if they left their pits without permission, even if they perceived their lives to be in imminent danger.[42] The bond also empowered mine owners to enforce strict discipline of attendance and extend surveillance over workers during sickness. In their 1825 list of grievances, *A Voice from the Coalmines*, Tyne and Wear pitmen complained that they were fined 2 s 6d for absence 'unless we produce a certificate from the surgeon certifying we are ill'.[43] In south Wales, colliers in the mid-nineteenth century were usually employed on monthly contracts.[44] At the other end of the spectrum were some mineworkers in Scotland. In Midlothian, nineteenth-century miners were bound to coal owners under similar terms to miners in north-east England, sometimes for durations of anything up to five years. Long as this was in relation to the contracts governing mineworkers in other parts of Britain, this was a short time compared to the *lifetime* bondage and virtual serfdom experienced by colliers in Scotland in previous centuries. The trend over the nineteenth century in the coal industry, however, was generally towards shorter-term contracts more likely to last months or weeks than years.[45]

The massive and rapid expansion of mining in the nineteenth century therefore increased the demand for mine labour and opened up many new opportunities that mineworkers were keen to exploit. Nevertheless, it also presented dangers of death and disablement that coloured public perceptions of coalmining and miners' own views of their occupation. The remaining sections of this chapter explore how economic expansion increased the risk of accidents and how the nature and conditions of mine work outlined so far shaped the experiences of miners with injuries and impairments and influenced their ability to return to work.

Accidents and disability in the mining industry

Fatal mining disasters exercised a powerful grip on the public imagination in nineteenth-century Britain and have dominated discussions of health and safety in historical studies of coalmining.[46] In the period 1835–1880, fourteen major disasters each claiming more than 100 lives occurred at British collieries. The deadliest incident among miners of south Wales, Scotland and north-east England at this time was the explosion that killed 207 people at Blantyre Colliery in Lanarkshire in October 1877 – the worst disaster in Scottish mining history.[47] In 1850 the risk of mortality by occupational accident was around four or five times higher for mineworkers (4.5 to 5.0 per 1000 employed per year) than for the rest of the working population.[48] Deadly explosions, in particular, were widely reported in the press and were a focal point for public concerns about safety in the workplace. They prompted investigations into mine safety by engineers, stimulated collective action by workers around the questions of risk and compensation for victims and their families and, ultimately, acted as a spur to increasing government regulation.[49]

Despite the preoccupation with fatalities, death was not the only outcome of mining accidents.[50] However, due to poor record keeping, the exact scale of non-fatal accidents is difficult to measure. Although the first public investigation into mining accidents conducted in 1835 had called for the future collection of accident statistics, this recommendation was not supported by legislation, and casualty reporting remained patchy and incomplete until much later in the century.[51] Nevertheless, the impression that emerges from the available sources suggests that non-fatal accidents were more common than fatal ones. In the early 1840s, for example, it was estimated that the ratio of accidents in south Wales with 'disabling' consequences to numbers employed was 1 to 500, compared to a ratio of fatal accidents to numbers employed of 1 to 1000.[52] In his report on the mines of the south Durham coalfield, Dr James Mitchell gathered data from three unnamed collieries detailing the 'time, place and mode of maiming' experienced by workers and the amount of time they had lost through sickness. Injuries included bruising, lacerations and fractures, and the more general description of being 'lamed', which could indicate both temporary and permanent impairment.[53] For the most part, though, mine owners and colliery doctors were unable – or unwilling – to provide information about injuries. Charles Forrest, surgeon of Hirwaun Iron Works, regretted that he could not furnish statistics 'touching the number of accidents and proportion of able-bodied men as compared with the disabled' since he had 'never kept a record of all the accidents occurring in the works, merely making notes of remarkable cases professionally'.[54] John Roby Leifchild remarked in his report on Northumberland and North Durham that 'accidents terminating short of death are seldom heard of beyond the place of their occurrence'.[55] In general, then, official reports during most of the nineteenth century probably underestimated the number of non-fatal accidents.[56]

The reporting of non-fatal accidents did improve, however, as the century progressed, due to the increasing involvement of legislators in the regulation of industry. The Mines Act of 1850, for example, introduced a system of official inspection in the hope of reducing fatalities from explosions, and from 1855 mine owners and their agents had a legal responsibility to report 'any serious personal Injury' to inspectors.[57] While at first this was interpreted as referring primarily to injury by explosion, by the time of the 1872 Mines Act, the requirement had been extended to the reporting of serious injury 'by any accident whatever'. Failure to report serious accidents, moreover, became a punishable offence, with fines levied potentially going to workers who had been hurt in mining.[58]

Reporting of 'serious injury', however, was highly variable. As J. B. Atkinson, the inspector for South Durham noted in his report for 1863, 'a considerable amount of doubt, as well as difference of opinion, appears to prevail amongst the agents of the collieries of the district as to the amount of personal injury which constitutes what the Inspection Act terms serious personal injury'.[59] Reporting 221 non-fatal accidents in 1862 and 239 in 1863, the inspector for west Scotland's mines cautioned that 'it would be fallacious to accept these [figures] as a correct and full return of accidents which happen in mines,' since the term 'serious personal injury' is 'difficult to determine'.[60] Although this may have led to the underreporting of non-fatal accidents, in some cases mine owners were anxious to report any incident to make sure they stayed on the right side of the law. This was particularly true after 1872, when the law left no 'doubt' about mines' legal responsibility to 'notify' the authorities about all 'death arising from slight injuries'.[61]

The lack of comprehensive health and safety statistics for our period makes it hard to discern the relationship between industrial change and the incidence of impairment with any great precision.[62] For many eighteenth- and nineteenth-century Britons, however, a link between the expansion of the coal industry and a presumed increase in the occupational risks facing miners seemed clear. While the comparatively small size of pits in the early modern period had tended to keep accidents and fatalities to relatively low levels, the increasing scale and depth of mines from the late eighteenth century onwards, along with the rapid growth in the number of workers employed in the sector, began to raise concerns about the safety of coal pits.[63] These fears were fed by growing and sensational reports of accidents in and around coal mines. In the late eighteenth century, such reporting was a regular, if infrequent, feature of press coverage of 'casualties'. For example, in 1788 the *Whitehall Evening Post* reported an accident at Stratton in Somerset, when two mineworkers were killed and 'six more burnt in a terrible manner' after the 'damps of the pit took fire'.[64] As demand for coal increased and mines were driven deeper, more 'gassy' and dangerous seams were worked, increasing the risk of explosions. By 1813, worries about the number of underground explosions in north-east England led to calls there for the establishment of the Society for the Prevention of Accidents in Coal Mines, motivated by the 'humane purpose' of saving the lives of those involved in an industry that 'contribute[s] so essentially to our comforts'.[65] In the opinion of colliery viewer John Buddle, giving evidence to the House of Lords in 1829, it was those collieries in north-east England 'producing the best coals' that were most liable to explosions that caused loss of life or left an 'immense number of cripples'.[66] With increasing profits available as demand for coal rose, moreover, miners were exposed to new risks alongside the threat of explosions. For example, the practice of 'robbing' supporting pillars of as much coal as possible to maximise output not only required miners to work in extremely gassy parts of mines, it also meant they were at greater risk of roof falls. In addition to these more dramatic risks, deeper and more intensive mining also meant miners had to work in hotter and more stuffy environments, to the further detriment of their health.[67]

Deep mining necessitated innovations in ventilation, drainage, winding gear, lighting and transportation.[68] Ventilation, in particular, became a critical concern, as public worries about the increasing frequency of explosions mounted. New methods of ensuring the flow of air through underground passages were developed. These included the use of furnaces at the pit bottom to draw air through the upcast shaft. By 1830, such ventilation techniques were commonly employed at British collieries. Further efforts to improve ventilation were made from the 1840s onwards, when exploitation of the explosion-prone steam-coal seams of south Wales intensified. Experiments with air pumps and mechanical fans, for instance, were conducted.[69] Together with the system of government inspection, these developments probably made mines safer. Fatalities from explosions fell during the second half of the nineteenth century and former hewer Thomas Burt certainly believed that health and safety at British mines had improved alongside industrial expansion in the nineteenth century. Born in 1837, Burt came from a long line of miners. 'My paternal grandfather, who died before I was born,' recalled Burt in his autobiography, experienced debilitating respiratory problems brought on by a life in mining. In Burt's view, his grandfather's 'early death was not improbably due to the bad underground ventilation of those days'.[70] However, the exposure to risk and injury varied between and within mining districts. Such differences stemmed from the different geological conditions and extraction methods in the coalfields, as well as the uneven pace of economic development in the industry.[71]

Although technological and regulatory developments had beneficial effects, they did not eradicate the considerable risks faced by colliery workers. Sometimes, in fact, technological advancements actually posed new dangers for the workforce.[72] In the eighteenth century, water-powered machinery was often used at collieries for winding and pumping and was regarded as an improvement on human or animal powered technologies.[73] Yet such machinery could prove hazardous to use. In June 1787, the *Public Advertiser* reported that one man was fatally mangled when his clothes became caught in a chain attached to a colliery waterwheel near Oswestry. In attempting to save the poor man, another person's hand was also severely 'crushed' in the incident.[74]

The introduction of the miner's safety lamp (popularly known as the Davy lamp after its inventor, Humphrey Davy) around 1815 reduced the risk of explosion from naked flames, doubtless saving lives, but its use was controversial and varied between coalfields. Safety lamps gave off poor light, which made some miners abandon them, sometimes with tragic results. In *A Voice from the Coal Mines* (1825), colliers from Tyne and Wear complained that safety lamps were a great advantage to coal owners 'but a great injury to the comfort and earnings of the pitmen'. Coal owners fined workers for any deficiency in separating inferior coal from the superior, but this proved difficult in the dim light provided by the Davy lamp. Furthermore, they argued, the Davy lamp was an 'accessary to the destruction of our health; bringing on rapid old age and general imbecility', wrecking the eyesight of miners and encouraging mine owners to take risks in forcing their workers to mine more dangerous places.[75]

Unionised miners may have overstated the nature of the risks they faced to convince the public of the justness of their industrial disputes with owners. Yet others broadly agreed with them, including arguably the most famous mining expert of the time, John Buddle. Four years after publication of *A Voice from the Coal Mines*, Buddle echoed the claims of pitmen when he gave his opinion on the effects of the Davy lamp on mining safety in north-east England to Parliament. While Buddle accepted that safety lamps had probably reduced the number of explosions, he did not believe this had radically improved mine safety. The adoption of this new technology, in his view, had merely changed the risks to which miners were exposed. Comparing fatality statistics before and after the introduction of Davy lamps, Buddle reckoned 'the loss of life has been ... about the same'. The reason for this was that 'we are [now] working mines, from having the advantage of the safety lamp, which we could not have possibly worked without it, and of course they are in a more dangerous situation, and the risk is increased in a very great degree'.[76]

No matter what safety regulations or technological improvements were introduced at mines, however, it was impossible to eradicate the folly of individual workers. The 'rush and recklessness' that Thomas Burt witnessed at Murton Colliery was believed by many to be endemic among nineteenth-century mineworkers, and such claims were sometimes used to undermine pitmen's claims to occupational status.[77] The opinion was reinforced in the press and in the reports of government officials. In Scotland in 1863 it was reported that many 'inexperienced and rash men' were the cause of underground accidents.[78] In 'A Coal Miner's Evidence', which purported to be a first-hand account of an explosion from a Durham miner published in *Household Words*, it was claimed that many mining deaths were caused by the 'carelessness and folly' of miners: 'It's just in our nature not to care – that's all.'[79]

Perhaps some miners were just 'careless' by nature, but the class prejudice apparent in many reports about the 'reckless' behaviour of industrial workers ought to be recognised too. Hewers took risks with their health and safety not simply because they were foolish, but because they sought to maximise their earnings in an incredibly dangerous industry. As noted earlier, the payment of hewers by piece rate may have encouraged some to take extra risks with their lives and health for the sake of an increased pay packet. Seen in this light, then, colliers appear more as rational, if sometimes fallible, actors taking calculated risks than unthinking fools.[80]

While explosions were the focus of much discussion of safety in mines, inspectors' reports highlight other dangers, particularly roof falls and accidents relating to the transport of coal, as the primary causes of non-fatal injuries.[81] For instance, Lionel Brough, whose inspection district covered the south-west of England and part of south Wales, reported seventy-five non-fatal accidents involving ninety-one persons in 1863, in which there were twice as many accidents caused by falls of coal and stone (thirty) than those caused by explosion of 'fire damp' (fifteen). Thirty-four men and boys suffered fractures, burns and other injuries from accidents in shafts or 'by miscellaneous causes'.[82] Brough's report for the year 1869 listed 125 persons injured in 111 accidents, of which the largest number (sixty) were 'wounded, contused, or suffered from fracture of bone, or underwent amputation' from falls of coal or stone.[83] The report for the west of Scotland similarly listed falls of rock or coal as the primary cause of non-fatal injury, accounting for ninety-nine casualties.[84] In the report for north-east England that year, twenty-four workers were injured by falls of stone or coal, while eleven were 'run over or crushed by tubs'.[85]

The bodily risks mineworkers were exposed to varied according to the kind of tasks they were expected to perform. Consequently, the perils of mining did not affect all colliery workers equally. Putters or trappers, for instance, were more likely to be injured in accidents involving coal tubs than hewers. Given that a mineworker's position in the underground workforce was often directly related to his age, moreover, it seems age and experience were important determinants of risk. As Jamie Bronstein has observed, nineteenth-century fatality figures suggest that young and inexperienced mineworkers were more likely to be killed or injured at work than older, more experienced, colleagues.[86]

In general, the 'serious injuries' reported by the mines inspectors did not make clear distinctions between those which threatened a worker's livelihood through permanent disablement and those from which recovery was possible, nor did they take account of other long-term health problems such as respiration difficulties caused by inhaling dust or physical problems caused by working in damp, cramped environments. Though inspector Herbert Mackworth estimated in his report for south Wales and south-west England for 1855 that about twice as many were 'disabled for life' by mine accidents than were killed, inspectors noted that many victims of 'serious injury' were 'restored to usefulness'.[87] Of the seventy-five accidents reported to Hedley in 1862 that injured eighty persons, 'only one of these ... has resulted in a permanent deformity' – a man who was seriously hurt by leaning out of the cage as he descended the mine shaft, which resulted in facial disfigurement and the loss of sight in one eye.[88] It is unclear whether this man's 'permanent deformity' equated to an inability to work. The nature of the injuries sustained in accidents and their effects on individuals varied considerably. For example, among the victims of falls of rock or coal listed in the East Scotland Mines Inspection Report for 1880 were a twenty-six-year-old collier whose leg was injured resulting in just seven days off work and an eighteen-year-old 'brusher' whose arm and leg were broken, leading to a lay off of 120 days. A drawer from the Arniston Coal Company's mine aged fourteen whose legs were 'injured by chain on a wheel brae' had been unable to work for eight months and was 'still off'.[89] The final section of this chapter examines the experiences of injured or impaired miners in more detail to reassess the relationship between 'disability' and the industrial workplace.

Disability and work

Many factors influenced the perceptions and experiences of mineworkers with impairments. The nature of their bodily capacities was undoubtedly significant in this regard, but so too were the conditions and organisation of work set out earlier on in this chapter. These influenced the 'somatic flexibility' available in the industrial workplace and affected the ability of workers with impairments to participate in the labour force. The attitudes of employers and fellow workers were also important, as these could determine whether impaired mineworkers were actually welcome at mines or not.

In the eighteenth and nineteenth centuries it was common to define disability in relation to work, but impairment was a matter of degree and did not necessarily equate to an inability to do any work at all – something that, as we will see, welfare officials at the time were often keen to stress.[90] In line with other workers, mineworkers could potentially experience two kinds of disability: *occupational* and *general* disability.[91] As the term implies, occupational disability meant that a person was unable to pursue their normal employment due to ill health or injury. Referring to the Scottish town of Tranent in 1840, which had a high number of infirm, ill and injured colliers among its population, sanitary inspector S. Scott Alison relied on an occupational definition of disability when he wrote:

> A great number of persons in and around Tranent are unfit to follow their *usual occupation* on account of bodily injuries by accidents, and of disease caused by their occupation.[92]

While this meant that people were disabled from doing their usual work it did not mean that they were entirely incapable of labour. *General* disability, in contrast, meant a person was totally incapacitated for *any* kind of work. This definition of disability was perhaps the most dominant at the time. As Martha Stoddard Holmes contends, the idea that disabled people were incapable of work was an incredibly popular one in Victorian literature and popular culture.[93] Despite such popular representations, however, the relationship between 'disability' and inability to 'work' was rarely straightforward in practice.

The history of Britain's coalfields provides compelling evidence of this complexity. It was not inevitable that illness and injury forced mineworkers from colliery labour forces. As is clear from the description of men with crutches and wooden legs encountered by Thomas Burt working in the mines of mid-Victorian Durham that began this chapter, people with quite significant impairments were not excluded from underground work. Indeed, the 1842 Children's Employment Commission had returned evidence of many people with 'disabilities' working in Britain's coal mines. Among those working at Birsley Colliery near Tranent were William M'Neil, an eleven-year-old boy, 'deaf and dumb, who had wrought below two years'. At Polkemmet Colliery in Linlithgowshire (West Lothian) worked another eleven-year-old, Catherine Thomson, who had returned to work after having her knee crushed by a cart, which continued to cause her great pain and she could 'scarcely stand' after pushing carts all day. Another child worker, Taylor Coats, who hooked and unhooked chains off rolleys at Percy Main Colliery in Northumberland, had been 'lamed twice' by his work, resulting in absences of three weeks and twenty-two weeks respectively. Now he 'walks lame. Has a bad step; cannot walk comfortably' but continued his work at the pit.[94] Undoubtedly, the evidence of impaired children working in the mines served to highlight the horrors of the industry, just as reformers had used accounts of young 'factory cripples' in the previous decade to campaign for regulation of the manufacturing industries.[95] Nevertheless, these accounts served to reinforce the view that injury, impairment and living with pain or other difficulties was part of the experience of work in the coal industry. 'Working through' sickness or impairment was part of nineteenth-century British working-class experience and was not simply a rhetorical image used by reformers to garner support for their cause.[96] 'Disabled' Britons worked throughout the nineteenth century, often in some of the most physically arduous industries of the time.

The most vivid first-hand account of the working life of an impaired miner is that left by trade unionist Edward Rymer. Born in Boldon near Sunderland in 1835, Rymer was seriously hurt in a house fire as a young child, which left him significantly visually impaired and 'permanently injured' on the right side of his body. His defective eye-sight had, he wrote, a 'sad and serious effect on my whole life', and he referred to himself as a 'cripple'. Nevertheless, he was employed in a variety of jobs underground in pits in the north-east and other parts of the country, beginning work like many young people did as a trapper, moving on to hauling tubs of coal as a putter and eventually, in 1860, to hewing coal. At each stage of his career, Rymer wrote of the ways in which his impairments, and the attitudes of others to them, affected his experience of work. If his 'defective sight' did not act as a barrier to mine work, he wrote frequently of how it made his work more onerous, or caused others to exploit him. It caused him 'endless troubles' as a young trapper, leading those of his fellow miners who were 'disposed to act the part of tyrant or persecutor' to impose extra tasks upon him. He felt isolated through his impairment, describing himself as a 'lad half blind and friendless' and, when working as a horse-driver, wrote that 'my failing sight frequently led me to make mistakes'. It was at this point, he wrote, that his 'partial blindness ... dawned on me with full force, and brought out all the cunning in my nature, since I had to watch every point, crossing, and turn in the workings'. The physical exertion also took a toll on his limbs, as did working in damp parts of the mine, which brought on rheumatism. Fearing physical 'ruination' when he was assigned to a haulage job after having worked for a time as a hewer, Rymer asked to return to coal-cutting, but was refused. His preference for hewing was so strong that he broke his bond and absconded – an offence for which he was later imprisoned. Rymer's wish to work as a hewer was probably affected by the better pay and status it attracted, but he also seems to have thought that hewing offered less severe challenges to his body than hauling coal through dark passages.[97]

While the high status and better pay associated with coal-cutting may have made it the job of choice for many mineworkers, not all colliers were able to resume their occupation after injury. Some miners were so badly injured that they were unable to muster the strength or bodily dexterity necessary to make a living as hewers. This did not always stop them working at lighter jobs however. As we have seen, the coal industry was occupationally diverse. In 1829, John Buddle explained that in north-east England men made 'cripples' in mine accidents needed to be 'provided with some employment which they could manage'. Consequently, 'many of them go to boys' work, what is called trapping, furnace keeping, and a great many jobs that might be done by boys, if it was not for the sake of employing these cripples and disabled persons'.[98] Some impaired miners took up surface work. Evidence given by William Morrow, a fourteen-year-old employee of Tyne Main Colliery, to the 1842 Children's Employment Commission provides a vivid example of how an individual's working life might change with injury. William had begun work underground at a different pit as a trapper, but had fallen asleep one night by his door and some wagons ran over him, shattering his leg. The leg was amputated and he was fitted with a prosthetic limb, but he was able to return to his door keeping work. Nevertheless a series of accidents, including having his head cut by a fall of rock, breaking his arm and cutting his brow when coal fell from a wagon, had led to him taking work on the surface, where he kept the 'foot of the inclined bank'.[99] Old and infirm colliers often worked aboveground as 'callers' or 'knockers' who roused mining households to ensure workers were up in time for their shifts.[100] Others became 'lampmen' and were responsible for checking, maintaining and storing safety lamps.[101] Injured mineworkers were not automatically cast off and excluded from colliery work. Many were able to continue working in the coal industry, both above ground and below.[102] Some resumed their pre-injury occupations while others took up new, less arduous positions.

The status of miners with disabilities varied considerably and might depend on a person's standing prior to their injury and the work culture of particular collieries. Those like Rymer whose impairments were acquired prior to working in the mines sometimes faced the scorn or suspicion of co-workers. Rymer frequently complained of his physical defects being used against him, such as in a dispute with an overman over a fair price for putters' labour in which his opponent 'made some brutal remarks' about his 'lameness'.[103] The assignment of hewers to 'boys' work' might have enabled them to return to work after serious injury, but in an industry where there was a strict occupational hierarchy based on age, such tasks might be deemed demeaning as they would almost certainly carry a loss of earnings and status.

In contrast, as Edward Slavishak has shown, a certain degree of bodily marking or minor impairment might confer status on workers in nineteenth-century heavy industry as signs of experience or bravery.[104] In an industry where some degree of 'seasoning' of workers' bodies was expected as part of the process of training and preparation for the arduous nature of pit life, a miner's bodily characteristics instantly placed him in the hierarchy of age and experience that characterised underground work.[105] The backs of putters, for instance, were often scarred as a result of their work. Peter Rutter, a fifteen-year-old youth who worked at Gosforth Colliery in the 1840s, explained that pushing heavy corves of coal through narrow passages meant 'the skin of his back is often knocked off'. At Monkwearmouth, an exceptionally hot mine, workers commonly experienced painful boils when they first went underground, but these tended to disappear over time. Consequently, 'the occurrence of these [boils] is so distinctive a mark of a fresh man, that [an observer] is well aware of the man being unaccustomed to the pit from this occurrence'.[106] Indeed, Victorian expert on occupational health J. T. Arlidge described certain 'deformities' associated with particular industries as 'trade-marks', which acted as badges of occupational identity.[107]

Loss of status or emasculation in the workplace as a result of injury, then, was possible but not inevitable.[108] Although some men may have been disabled from returning to their former employment, their knowledge and experience might have made them suitable for supervisory roles. Thomas Haswell, killed in an underground explosion at Thornley Colliery near Sunderland in 1841, was an experienced coal hewer, but had been working as an overman, supervising the preparation of the mine between shifts, after breaking both legs in a rock fall the year before. Though newspaper reports of the disaster described him as a 'cripple', Haswell's impairments may have marked him out as a survivor and a man with hard-won experience of mining, earning the respect of the young crew who worked for him.[109] Colliery overmen might be paid up to £100 a year in Haswell's day and were generally promoted from the workforce on account of their 'activity, steadiness, natural abilities and education'.[110] Even when completely incapacitated for work, experienced mineworkers could still occasionally play an active role in the working life of a colliery. Reese Price, for example, was also an overman, at Gethin Colliery in south Wales. While off work following an accident in December 1865, Price was visited by his replacement and asked for advice about a build-up of gas in the mine. Although 'unable to follow' his usual employment, then, Price remained influential.[111] His knowledge and experience of mining meant he was not pushed aside and forgotten. As we will see in Chapter 5, experience of working in mines was a marketable commodity that may have helped some injured mineworkers find work in the coal industry, particularly at times when experienced mining labour was in short supply, such as during strikes.

Former mineworkers disabled from working underground often found employment in other sectors of the coalfield economy too. For example, a 'disabled man' might occasionally 'take to dealing in small wares' in pit villages, according to one mid-century investigation into life in the mining districts of north-east England, 'and in some cases superannuated hewers will, in addition to what light jobs they can pick up about the pit, occupy themselves in collecting the clay used for the workmen's candles'.[112] Sometimes men disabled from working in the coal mines turned to teaching. During a parliamentary debate on education in Wales in 1846, it was revealed that eleven of Merthyr Tydfil's schoolmasters had been 'miners, or labourers disabled by accidents or bad health'.[113] Similarly, Leifchild wrote that in the pit villages of north-east England many teachers were 'disabled workmen'. The practice of appointing such men was, in Leifchild's view, evidence of sentiment and a sense of entitlement based on physical sacrifice triumphing over reason since few had 'received an education qualifying them to be successful teachers of the young'. The disabled miner claimed 'from the mutilation or absence of a limb, to be the recipient of the bounty of the benevolent':

> It is not seldom that a disabled pitman proposes himself as a candidate for a schoolmastership; and should a vacancy occur in the colliery where his misfortune happened, he deems himself, and is deemed by others, indisputably entitled to the suffrages of all parties. This evinces sensibility to the claims of misfortune, and insensibility to the claims of education.[114]

While for critics such as Leifchild the entry of disabled miners into village schools was evidence of the poor educational provision in mining communities, the belief that injured miners had a moral claim to a job that carried authority and status is interesting and provides further evidence that those disabled from mine work were not necessarily marginalised in this period. Although opportunities for disabled miners may have declined at times of economic hardship, such as during periods of unemployment caused by the temporary or permanent closure of a mine, the occupational make-up of coalfield settlements was sufficiently diverse to offer some disabled colliers viable alternatives to mine work.[115]

The nature of their impairments was not the only factor affecting the ability of colliers to return to pit work after injury. The organisation and nature of mine work itself at this time facilitated the inclusion of 'disabled' miners in the mining workforce by permitting a degree of 'somatic flexibility' supposedly characteristic of the more 'inclusive' pre-industrial economy.[116] That it did, suggests that pre-industrial approaches to work were not obliterated in the nineteenth century but persisted in industrial settings too. During the late eighteenth and nineteenth centuries, coalmining developed in ways that resisted the conventional models of 'industrialisation' emphasised in accounts of the rise of disabling capitalism.[117] For example, coal-cutting throughout this period was a matter of individual hand labour that made it difficult to supervise workers closely and led to a degree of elasticity in working arrangements.[118]

In the first place, the continuance of piecework, particularly for hewers, gave mineworkers some leeway to decide their pace of work.[119] Such flexibility may have allowed miners to tailor their working practices and rhythms to minimise the restrictive effects of impairment – even if those unable to win as much coal as stronger hewers faced the prospect of lower pay. George Parkinson, for example, who worked in the mines of mid-nineteenth-century Durham, described one hewer called 'Old Joe' who was 'not physically strong, his earnings were small and his means were scanty'.[120] The independence of colliers at work was further promoted by the 'bord and pillar' method of extracting coal. This method of mining (also known as 'pillar and stall' or 'stoop and room' mining) consisted of driving a number of passages running parallel to each other from which coal was taken, leaving pillars of rock to support the roof. Although particularly suited to mines where seams were thick and roofs were typically poor – such as those found in Durham, parts of Scotland and the Swansea area of south Wales – bord-and-pillar working was the dominant method for much of the eighteenth and most of the nineteenth centuries.[121] As the idealised picture of a coal mine in Figure 2 suggests, colliers employed at pits using this kind of system, or variants of it, worked in individual bords or stalls that were separated from those of their colleagues. Working these areas in small groups, colliers were hard for pit managers to supervise. Coupled with the fact that poor lighting and the sprawling scale of many mines made supervision doubly difficult, miners were essentially left to their own devices and could effectively choose how they worked. If there was work discipline in the mines, argues John Williams in his study of south Wales, it was above all self-discipline.[122]

Assistance from fellow workers also facilitated the participation of people with impairments in the working life of collieries. Philip Lloyd, a collier at Waterloo Colliery in Monmouthshire, told the 1842 Children's Employment Commission that one of his daughters 'works in the mine for William Morgan who has lost his leg and cannot do much'.[123] The presence of kin was particularly important in helping impaired miners remain economically active. In south Wales and Scotland, miners commonly worked in family groups. Even when they did not, many mineworkers across Britain still had family members working close by them at collieries.[124] Due to the practice of sons following their fathers into the pits, different generations of impaired miners sometimes worked in the same place. Among the victims of an explosion at Gethin Colliery near Merthyr Tydfil in February 1862 were William Lewis, a forty-seven-year-old collier, and his eighteen-year-old son. The elder Lewis had been 'ruptured and confined to his bed' for sixteen weeks before the disaster, while his deceased son had 'lost his leg in the same pit' some time before. Both were

Figure 2 A coal mine: miners at work above and below ground. Wellcome Library, London/CC-BY 4.0.

working underground at the time of the explosion. Newspapers reported that the Lewises had two more sons who had not been involved, including another who had lost a leg in the pit nine months previously, but had been spared from the explosion because he was off work then as the family did not have enough food to pack for his lunch.[125] There were several other father-and-son teams who worked together in the same mine.[126] It is not clear whether it was to William Lewis or another man that the *Cardiff Times* referred on 21 February when it described two unnamed victims of the explosion as a 'poor collier and his son' who worked together because the 'man was delicate in health, and earned but little'.[127] Three years later, Griffith Ellis, who had a wooden leg, suffocated after another explosion at Gethin Colliery, alongside his brother David.[128] Tragic as these examples are, they suggest that the help of kin may have enabled impaired miners to carry on working.

Despite the general similarities between coalfields, there were variations that influenced the ease with which people with impairments could take up productive roles in the coal industry. In some places, coal was easier to work. In others, it was far more challenging. These differences help explain the cavilling system in north-east England in which hewers and putters drew lots to determine their workplaces underground. The purpose of allocating places in this way was to eliminate accusations of partiality or unfairness. A miner's place, then, was down to chance, not the whim of a mine manager. This was of the utmost importance to Durham and Northumberland pitmen, as a miner's place, or 'cavil', could radically affect his productivity and therefore pay.[129] For 'crippled' workers, like Edward Rymer, the cavilling system sometimes worked to their advantage by giving them an easy cavil that allowed them to earn a reasonable living despite their diminished physical capacities. At other times, it disadvantaged them. At his second cavilling in Houghton Pit, Rymer drew what he called a 'heavy "flat"' and found his strength 'insufficient to bear the task of pushing and lifting iron tubs 12 hours a day'.[130] When working as a hewer at Old Grange Colliery in partnership with his brother Jack (also described as a 'cripple'), with whom he split shifts, Rymer also found himself 'disabled', in a sense, by the rate of pay. As Rymer himself put it: at Old Grange there were 'good hewers around us, and we were handicapped beyond our strength to get anything like a living at 5s a score'. Trying to keep up, his hands became covered with blisters that eventually required medical attention and kept him off work for 'several weeks'.[131] The cavilling system was unique to north-east England, but the experiences of men like Rymer under it suggest how geological conditions and the allocation of work across the coalfields might influence the ability of impaired or injured miners to participate in coal production in sufficiently profitable ways.

Nineteenth-century colliers proudly proclaimed their 'independence'. However, the flexibility of working arrangements that allowed participation of men with a variety of physical impairments was gradually undermined as the period wore on.[132] The rise of 'industrial time' and the increasing pace of life brought on by the Industrial Revolution is often regarded as a key factor in the 'disabling' of people with impairments.[133] As pieceworkers, miners throughout the nineteenth century often worked more to task than time. In the 1840s, Benjamin Martin, a mineral agent for Penydarren Ironworks in south Wales, remarked of colliers there that 'we do not look after their time; they work any number of hours that they like themselves, we pay only for the quantity they send out'.[134] Under such circumstances, strict industrial timekeeping was clearly not very important, either to miners or managers. Yet changes eroding colliers' ability to determine their own hours of work did occur, albeit slowly and unevenly. In the early phases of industrialisation, mining operations had taken place close to the surface, which made it fairly easy for miners to enter and leave mines as they chose. With the growth of deep-mining, however, colliers became more dependent on winding machinery to access mines and were increasingly tied to particular times for ascending and descending that regulated the working day.[135] The frustration miners felt about the constraints winding equipment placed on their ability to come and go as they pleased is suggested by the tragic fate of Robert Moore. In August 1833, newspapers reported Moore's death after he fell down a 130-yard-deep mineshaft at Dry Clough Colliery in Lancashire, where he was employed. Moore was said to have lost his footing after becoming 'enraged' because he arrived for work 'too late to descend with the coal tub, which had just left the shaft'.[136] Moore was obviously not a man with complete freedom to decide his own hours of work.

By the time of the 1842 Children's Employment Commission, miners in the west of Scotland were tending to work standardised twelve hour days that began at six o'clock in the morning and were 'anxious to get coal picked out in time to supply the engine' that began its daily task of lifting coal to the surface at six in the evening.[137] Although miners in different coalfields continued to work varying hours, the growing calls to reduce the working day to eight hours after 1850 may also be seen as a sign that miners, like other industrial workers, were increasingly subjected to time discipline.[138] The introduction of new forms of work discipline was particularly evident in the new integrated coal and iron companies that came to dominate coal production in the west of Scotland and parts of south Wales between the 1830s and 1870s. Production in these companies was geared to very different rhythms than those of smaller concerns supplying local domestic consumption, and the consistent rather than fluctuating demand for coal put greater pressure on worker productivity.[139] Taken together, these developments potentially reduced the 'somatic flexibility' available to impaired workers in the industry by undermining their ability to determine their own hours and patterns of work.

Changes in coal-extraction techniques may also have challenged miners' sense of independence and placed greater constraints on older or impaired workers working underground. Although bord-and-pillar mining was never fully displaced during the nineteenth century, from the 1860s 'longwall' mining became more popular, first in Scotland and later in south Wales and north-east England.[140] Whereas bord-and-pillar extraction favoured hewers working individually, longwall cutting revolved around teams of miners hewing coal from the seam alongside each other in a line. The system had economic advantages, particularly in increasing the volume of marketable coal and reducing waste.[141] This may have placed those unable to work at the same pace as others at a disadvantage and cast them as a risk to productivity or safety. James Thain, who worked in a team of longwall miners in Pembrokeshire at the end of the nineteenth century, recalled that those men who lagged behind made the working face 'irregular, which often caused the roof to break and the coal to become set or much harder to dig'. This was the cause of many arguments that required intervention from the manager or his assistant. Thain remembered hewing alongside an 'old man' who had a 'bad habit of allowing the part of the working face near to me to hang behind'. On at least one occasion, this led to a roof fall that trapped the old man. Luckily for him, the old worker survived the accident but was severely admonished by Thain who told him to take it as a lesson 'and don't let your place hang back again endangering the life of yourself and others'.[142] Thain's memory of the incident indicates that not all old or impaired mineworkers were always treated with kindness or sympathy by workmates. At times, their diminished capacities could also make them the target of discrimination, prejudice and hostility in the workplace.

The demise of stalls may also have increased managerial pressure on colliers. Although miners with impairments continued to work at collieries employing the longwall method, the greater coordination of workers required by this system meant they were more likely to come under close supervision than in bord-and-pillar mining. Keen to avoid disruptions to production like the one Thain described, it is possible that managers in charge of longwall mines became less tolerant of 'disabled' miners unable to 'keep up' and were less likely to assign them face work. Ultimately, longwall mining would lead to the greater adoption of machine cutting and the mechanisation of coal extraction in the twentieth century, further increasing the pace of mine work.[143] We should not, however, exaggerate the impact of these changes. The earning power of older miners remained healthy for most of this period, beginning to fall only in a man's fifties. Furthermore, even if their workmates did not particularly like it, 'old' men like the one encountered by Thain were still working at longwall mining well into the late nineteenth century and beyond.[144] Nevertheless, as we shall see in Chapter 5, the employment of older miners did become more contested towards the end of the nineteenth century, due to fears that they posed a financial risk to employers because of their perceived greater susceptibility to compensable injury. Changes in coal extraction may have added to these concerns.[145]

Conclusion

In contrast to the view that industrial expansion rapidly and decisively excluded impaired people from the workforce, evidence from Britain's coal industry in the period 1780–1880 indicates a more complicated picture in which experiences of 'disabled' miners varied considerably. The shift to coal-powered technology fuelled and sustained Britain's economic expansion in this period and contributed to the enormous growth of the coal industry. Yet the development of the coal industry was uneven and had distinctive regional characteristics. Systems of labour organisation and methods and conditions of work varied considerably, not only between coalfields, but also between individual collieries in each coalfield. As Thomas Burt's comments that began this chapter suggest, physical impairments were a common sight in Britain's nineteenth-century coal mines, but they were more visible in some mines than others. This reflected not only varying safety standards between collieries, which the system of national inspection brought in after 1850 sought to address, but also the local factors that permitted miners with impairments to re-enter the workforce after injury. In the case of women and young children, it was legislation prompted by moral concerns that played a more decisive role in 'disabling' them from working underground than physical incapacity.

The expansion of the coal industry was widely associated with growing risks to mineworkers, but the exact scale of injury or disablement is difficult to document in an era where reporting of non-fatal accidents was patchy and where there was little consensus on what constituted a 'serious injury'. It is therefore difficult to substantiate the view that industrial expansion produced impairment on a mass scale, although this idea was frequently used to critique industrial practices. There was no simple correlation between 'serious injury' and disability; on the one hand, mine inspector reports often stated that the seriously hurt were returned to 'usefulness', on the other, as we shall see in the following chapter, serious injury was not the only source of impairment for coalminers. Statistics for non-fatal injuries therefore provide limited evidence of the scale or nature of impairments facing mineworkers.

Taken together, the material examined in this chapter cautions against monolithic interpretations of disability and suggests there was no simple correlation between impairment and inability to work. During the so-called 'classical' period of Britain's industrial expansion, from the mid-eighteenth to the mid-nineteenth centuries, people with impairments continued to find work in mining. The survival of 'pre-industrial' working practices, such as the relative independence of hewers, the continuation of work in kin groups, and working to task, may have helped miners with impairments to remain economically active in the industry, retaining some of the 'somatic flexibility' characteristic of earlier periods.[146] Nevertheless, this may have been undermined over time by economic developments such as the spread of longwall mining and the stricter discipline imposed by mechanisation and the incessant demands of the iron industry in parts of south Wales and west Scotland.

This chapter has indicated the importance of understanding the nature and conditions of work and the culture and structure of the workplace to interpreting experiences of disability and industrialisation. A certain degree of impairment or physical scarring was expected and might mark a man out as a 'survivor' and earn him the respect of others. However, Edward Rymer and James Thain's memories of mining show how 'cripple' or elderly miners might face abuse. Their writings also illustrate the ways in which the structure of the workplace, including local customs of labour such as cavilling in north-east England, might variously work to enable injured mineworkers to earn a living or present disabling barriers. The work undertaken by miners with impairments was not always well paid, nor did the 'boys' work' done by some disabled men carry the same prestige as other tasks. Mines offered diverse employment opportunities, but the meanings of work adhered to a strict hierarchy of status, closely related to age. That said, the perception of some disabled people's work as 'lowly' does not mean it did not have value to those who undertook it. For Rymer, work for a 'cripple' such as himself was part of his 'hard fight to live', to support himself and avoid dependency and was a source of pride.[147] The employment of miners with impairments was consistent with cultural values that demanded that people with disabilities avoid becoming a 'burden' by remaining productive where possible. As we shall see in the following chapter, the whole edifice of medical care in Britain's coalfields was built upon this objective.

Notes

1 Thomas Burt and Aaron Watson, *Thomas Burt, M.P., D.C.L., Pitman and Privy-Councillor: an Autobiography, with Supplementary Chapters by Aaron Watson* (London: Unwin, 1924), 84, 93, 94.
2 Daniel Blackie, 'Disability and Work during the Industrial Revolution in Britain', in Michael Rembis, Catherine Kudlick and Kim E. Nielsen (eds), *The Oxford Handbook of Disability History* (New York: Oxford University Press, forthcoming).
3 PP 1842 (380), *Commission for Inquiring into the Employment and Condition of Children in Mines and Manufactories. First Report of the Commissioners*, 24–5.
4 Angela V. John, *By the Sweat of Their Brow: Women Workers at Victorian Coal Mines* (London: Routledge and Kegan Paul, 1984), 21; J. H. Morris and L. J. Williams, *The South Wales Coal Industry 1841–1875* (Cardiff: University of Wales Press, 1958), 214.
5 John, *Sweat of Their Brow*, 24.
6 Ibid., 24–5.
7 Ibid., 24; Robert Duncan, *The Mineworkers* (Edinburgh: Birlinn, 2005), 35, 70.
8 Duncan, *The Mineworkers*, 80–2.
9 John, *Sweat of Their Brow*, 22; Christopher A. Whatley, 'A Caste Apart? Scottish Colliers, Work, Community and Culture in the Era of "Serfdom", c. 1606–1799', *Journal of the Scottish Labour History Society*, 26 (1991), 3–20.

10 5 & 6 Victoria Cap. XCIX *An Act to Prohibit the Employment of Women and Girls in Mines and Collieries, to Regulate the Employment of Boys, and to Make other Provisions Relating to Persons Working Therein*, 10 August 1842.
11 John, *Sweat of Their Brow*.
12 Roy Church, *The History of the British Coal Industry*, vol. 3: *1830–1913: Victorian Pre-Eminence* (Oxford: Clarendon Press, 1986), 191.
13 Some children lied about their age to work below ground. For an example, see Edmund Stonelake in A. Mòr O'Brien (ed.), *The Autobiography of Edmund Stonelake*, (Bridgend: D. Brown and Sons, 1981), 49.
14 PP 1842 (380), 30.
15 'The Collieries – No. 1', *Monthly Supplement of the Penny Magazine of the Society for the Diffusion of Useful Knowledge*, 28 February–1 March 1835, 123–4.
16 John Benson, *British Coalminers in the Nineteenth Century: A Social History* (Dublin: Gill and Macmillan, 1980), 28–30.
17 George Parkinson, *True Stories of Durham Pit-Life* (London: C. H. Kelly, 1912), 1.
18 Alan Campbell and Fred Reid, 'The Independent Collier in Scotland', in Royden Harrison (ed.), *Independent Collier: The Coal Miner as Archetypal Proletarian Reconsidered* (Hassocks: Harvester Press, 1978), 57; Robert Colls, *Pitmen of the Northern Coalfield: Work, Culture and Protest, 1790–1850* (Manchester: Manchester University Press, 1987), 11, 15.
19 Parkinson, *True Stories*, 1.
20 'On the Habits and Diseases of Northern Pitmen', *Colliery Guardian*, 12 September 1863, 204.
21 'The Collieries', 127.
22 Herbert Mackworth, 'Mines: Accidents in Them, and Sanitary Condition of Them' in Bristol Mining School, *Lectures Delivered at the Bristol Mining School, 1857* (Bristol: Bristol Mining School, 1859), 183.
23 M. W. Flinn, *The History of the British Coal Industry*, vol. 2: *1700–1830: The Industrial Revolution* (Oxford: Clarendon Press, 1984), 87, 91, 106–8.
24 Campbell and Reid, 'Independent Collier in Scotland', 57.
25 Morris and Williams, *South Wales Coal Industry*, 235.
26 Martin Daunton, 'Down the Pit: Work in the Great Northern and South Wales Coalfields, 1870–1914', *Economic History Review*, 34:4 (1981), 578–97; Morris and Williams, *South Wales Coal Industry*, 190–1.
27 Daunton, 'Down the Pit'; Morris and Williams, *South Wales Coal Industry*, 190–1.
28 Benson, *British Coalminers*, 54.
29 Anthony Errington, *Coals on Rails, Or the Reason of My Wrighting: The Autobiography of Anthony Errington, a Tyneside Colliery Waggon and Waggonway Wright from his Birth in 1778 to around 1825*. P. E. H. Hair (ed.) (Liverpool: Liverpool University Press, 1988); Benson, *British Coalminers*, 29.
30 Cited in Church, *British Coal Industry*, 195.
31 Burt, *Autobiography*, 109.
32 Daunton, 'Down the Pit', 590.

33 Stonelake, *Autobiography*, 55.
34 Rev. Thomas Gisborne, 'A General View of the Situation of the Mining Poor, Compared with That of Some Other Classes of the Poor' in Society for Bettering the Condition and Increasing the Comforts of the Poor, *The Reports of the Society for Bettering the Condition and Increasing the Comforts of the Poor*, vol. I (London: W. Bulmer and Co., 1798), 369.
35 Church, *British Coal Industry*, 571.
36 For example, PP 1842 (380), 157; cf. H. H. B., *Black Diamonds; or, the Gospel in a Colliery District* (London: James Herbert and Col, 1861), 123–4.
37 Steve Sturdy, 'The Industrial Body' in Roger Cooter and John Pickstone (eds), *Companion to Medicine in the Twentieth Century* (London: Routledge, 2003), 218.
38 Jamie L. Bronstein, *Caught in the Machinery: Workplace Accidents and Injured Workers in Nineteenth-Century Britain* (Stanford, CA: Stanford University Press, 2008), 158.
39 Benson, *British Coalminers*, 64; Colls, *Pitmen*, 53.
40 Benson, *British Coalminers*, 76–9.
41 Church, *British Coal Industry*, 576.
42 'The Collieries', 125; PP 1842 (380), 40; Huw Beynon and Terry Austrin, *Masters and Servants: Class and Patronage in the Making of a Labour Organisation; the Durham Miners and the English Political Tradition* (London: Rivers Oram, 1994), 29–46; Colls, *Pitmen*, 45–73; Bronstein, *Caught in the Machinery*, 120, 137.
43 United Association of Colliers, *A Voice from the Coalmines, Or, A Plain Statement of the Various Grievances of the Pitmen of the Tyne and Wear: Addressed to the Coal Owners, their Head Agents, and a Sympathizing Public* (South Shields: J. Clark, 1825), 25.
44 PP 1842 (380), 40; Morris and Williams, *South Wales Coal Industry*, 234.
45 John. A. Hassan, 'The Landed Estate, Paternalism and the Coal Industry in Midlothian, 1800–1880', *The Scottish Historical Review*, 59 (1980), 77. Despite Hassan's claim, bonds of five years were probably quite exceptional in nineteenth-century Scotland. In 1842, the Commission on Children's Employment reported that colliers in east Scotland were commonly hired on two-week contracts. PP 1842 (380), 40. Whatley, 'A Caste Apart?'
46 Bronstein, *Caught in the Machinery*, 11.
47 Observations here are based on table 4 in Benson, *British Coalminers*, 219; Duncan, *The Mineworkers*, 119.
48 P. E. H. Hair, 'Mortality from Violence in British Coal Mines, 1800–1850', *Economic History Review*, 21:3 (1968), 559.
49 Mills, *Regulating Health and Safety*, 80–4 and ch. 4 *passim*.
50 John Benson, 'Non-Fatal Coalmining Accidents', *Bulletin of the Society for the Study of Labour History*, 32 (1976), 20–2.
51 P. W. J. Bartrip and S. B. Burman, *The Wounded Soldiers of Industry: Industrial Compensation Policy 1833–1897* (Oxford: Clarendon Press, 1983), 21–2.
52 PP 1842 (380), 152.

53 PP 1842 (381), *Children's Employment Commission. Appendix to the First Report of the Commissioners. Mines. Part 1. Reports and Evidence from Sub-Commissioners*, 140.
54 PP 1842 (382), *Children's Employment Commission. Appendix to First Report of Commissioners. Mines. Part 2.* 553.
55 PP 1842 (381), 550.
56 Mills, *Regulating Health and Safety*, 19.
57 Ibid., 80–4; 18 & 19 Victoria Cap. CVIII, *An Act to Amend the Law for the Inspection of Coal Mines in Great Britain*, 14 August 1855.
58 35 & 36 Victoria Cap. LXXVI, *An Act to Consolidate and Amend the Acts Relating to the Regulation of Coal Mines and Certain Other Mines*, 10 August 1872. The question of compensation is explored more fully in Chapter 5. See also, Sturdy, 'Industrial Body', 220–1.
59 PP 1864 [3352] *Mines. Reports of the Inspectors of Mines, to Her Majesty's Secretary of State, for the Year 1863*, 62.
60 Ibid., 166.
61 PP 1875 [C.1216], *Mines. Reports of the Inspectors of Mines, to Her Majesty's Secretary of State, for the Year 1874*, 113.
62 Bartrip and Burman, *Wounded Soldiers of Industry*, 8–13.
63 Flinn, *British Coal Industry*, 412; cf. Mills, *Regulating Health and Safety*, 13.
64 *Whitehall Evening Post*, 14–17 June 1788.
65 Tyne and Wear Archives, S.PAM/1/6, Printed Address Urging the Formation of a Society for the Prevention of Accidents in Coal Mines, 1 September 1813.
66 PP *1830 (9), Report from the Select Committee of the House of Lords Appointed to Take into Consideration the State of the Coal Trade in the United Kingdom; with the Minutes of Evidence Taken Before the Committee, and an Appendix and Index*, 31.
67 Church, *British Coal Industry*, 322; Colls, *Pitmen*, 20–3.
68 Church, *British Coal Industry*, ch. 4 outlines these innovations in detail.
69 Ibid., 322–3.
70 Hair, 'Mortality from Violence', 545; Burt, *Autobiography*, 22.
71 Benson, *British Coalminers*, 41; Hair, 'Mortality from Violence', 557.
72 As Rose notes, historians of other countries or industries have also noted the negative impact of new technologies on the health and safety of industrial workers. Sarah F. Rose, '"Crippled" Hands: Disability in Labor and Working-Class History', *Labor*, 2:1 (2005), 35.
73 Stephen Hughes, *Collieries of Wales: Engineering and Architecture* (Aberystwyth: Royal Commission on the Ancient & Historical Monuments of Wales, 1994), 65–8; Flinn, *British Coal Industry*, 110–28.
74 *Public Advertiser*, 13 June 1787.
75 United Association of Colliers, *Voice from the Coal Mines*, 9, 12, 16; Mills, *Regulating Health and Safety*, 18.
76 PP 1830 (9), 32.
77 Colls, *Pitmen*, 15.

78 PP 1864 [3352], 158; Bronstein, *Caught in the Machinery*, 114.
79 'A Coal Miners Evidence', *Household Words*, 2:37 (1850), 245, 247.
80 Bronstein, *Caught in the Machinery*, 14.
81 Ibid., 11–12. For earlier examples see *Statistical Compendium*, table 2.5. Accidents and Injuries to Men and Boys Employed Underground in the Colliery of East Holywell, During the Year Ending May 24th 1841, provided by William Morrison, 'Medical Gentleman', to the 1842 Children's Employment Commission, http://doi.org/10.5281/zenodo.183686, accessed 24 March 2017.
82 PP 1864 [3352], 120.
83 PP 1870 [C.124], *Mines. Reports of the Inspectors of Mines, to Her Majesty's Secretary of State, for the Year 1869*, 84.
84 Ibid., 129.
85 Ibid., 142.
86 Bronstein, *Caught in the Machinery*, 13–14.
87 PP 1856 [2132], *Coal mines. Reports of the Inspectors of Coal Mines, to Her Majesty's Secretary of State*, 116; PP 1864 [3352], 120.
88 PP 1864 [3252], *Mines. Reports of the Inspectors of Mines, to Her Majesty's Secretary of State, for the Year 1862*, 55.
89 PP 1881 [C.2903], *Mines. Reports of the Inspectors of Mines, to Her Majesty's Secretary of State, for the Year 1880*, 196.
90 David M. Turner, *Disability in Eighteenth-Century England: Imagining Physical Impairment* (New York and London: Routledge, 2012), 127–30. On welfare, see ch. 3.
91 Blackie, 'Disability and Work'.
92 PP 1842 (008), *Sanitary Inquiry – Scotland, Reports on the Sanitary Condition of the Labouring Population of Scotland, in Consequence of an Inquiry Directed to be Made by the Poor Law Commissioners*, 123. Our emphasis.
93 Martha Stoddard Holmes, 'Working (with) the Rhetoric of Affliction: Autobiographical Narratives of Victorians with Physical Disabilities', in James C. Wilson and Cynthia Lewiecki-Wilson (eds), *Embodied Rhetorics: Disability in Language and Culture* (Carbondale and Edwardsville: Southern Illinois Press, 2001), 27.
94 PP 1842 (381), 470, 478, 577. For other examples, see Bronstein, *Caught in the Machinery*, 56.
95 Peter Kirby, *Child Workers and Industrial Health in Britain, 1780–1850* (Woodbridge: Boydell Press, 2013), 70–3.
96 James C. Riley, *Sick, Not Dead: the Health of British Workingmen during the Mortality Decline* (Baltimore, MD and London: Johns Hopkins University Press, 1997), 135.
97 Edward Rymer, *The Martyrdom of the Mine, or a Sixty Years Struggle for Life* (Middlesbrough, 1898), 2, 3, 5–7.
98 PP 1830 (9), 33.
99 PP 1842 (381), 631.

100 Benson, *British Coalminers*, 114; 'On the Habits and Diseases of Northern Pitmen,' *Colliery Guardian and Journal of the Coal and Iron Trades*, September 12, 1863, 204; Rymer, 'Martyrdom of the Mine', 4.
101 For example, see 'The Great Explosion at the Gethin Colliery. The Coroner's Inquest' (cutting from *Merthyr Telegraph*, Saturday, January 13, 1866) in TNA, HO 45/7729, Home Office: Registered Papers.
102 As Sarah Rose recognises when she cites the work of Alan Derickson, this also appears to have been the case at many American coal mines in the late-nineteenth and twentieth centuries. Rose, '"Crippled" Hands', 52.
103 Rymer, *Martyrdom of the Mine*, 8.
104 Edward Slavishak, *Bodies of Work: Civic Display and Labor in Industrial Pittsburgh* (Durham, NC and London: Duke University Press, 2008), 162.
105 Alan B. Campbell, *The Lanarkshire Miners: A Social History of their Unions, 1775–1974* (Edinburgh: John Donald, 1979), 39.
106 PP 1842 (381), 595, 645.
107 J. T. Arlidge, *The Hygiene, Diseases and Mortality of Occupations* (London: Percival and Co. 1892), 16.
108 Blackie, 'Disability and Work'.
109 'Explosion in a Coal-Pit, and Melancholy Loss of Life', *The Monmouthshire Merlin*, 14 August 1841.
110 PP 1842 (381), 128.
111 'The Great Explosion at the Gethin Colliery', *Merthyr Telegraph*, 13 January 1866.
112 Jules Ginswick (ed.), *Labour and the Poor in England and Wales, 1849–1851: the Letters to the Morning Chronicle from the Correspondents in the Manufacturing and Mining Districts, the Towns of Liverpool and Birmingham, and the Rural Districts*, 8 Vols (London: Frank Cass, 1983), ii, 63.
113 For other examples, see *Hansard*, HC Deb 10 March 1846, vol. 84, cols 847–8. See also *Merthyr Telegraph*, Saturday 29 May 1858.
114 J. R. Leifchild, *Our Coal and Our Coal-Pits* (London: Longman, Brown, Green, Longmans, 1855), 208, 214.
115 There is also evidence that disabled people in coalfield communities such as Shields on Tyneside may have made a living as water fountain attendants. See PP 1842 (007), *Sanitary Inquiry – England. Local Reports on the Sanitary Condition of the Labouring Population of England, in Consequence of an Inquiry Directed to be Made by the Poor Law Commissioners*, 443 (Greenhow's evidence).
116 Brendan Gleeson, *Geographies of Disability* (London: Routledge, 1999), 87.
117 Michael Oliver and Colin Barnes, *The New Politics of Disablement* (Basingstoke: Palgrave Macmillan, 2012), 61.
118 John Williams, *Was Wales Industrialised? Essays in Modern Welsh History* (Llandysul, Dyfed: Gomer Press, 1995), 27.
119 Benson, *British Coalminers*, 55–7; Campbell, *Lanarkshire Miners*, 35, 109.
120 Parkinson, *True Stories*, 35.
121 Flinn, *British Coal Industry*, 82–7, 90–9.

122 Daunton, 'Down the Pit', 583; Williams, *Was Wales Industrialised?*, 31; Colls, *Pitmen*, 29; Campbell, *Lanarkshire Miners*, 35.
123 PP 1842 (382), 536.
124 For examples, see Burt, *Autobiography*, 111; Parkinson, *True Stories*, 18–20.
125 *Cardiff Times*, 28 February 1862.
126 As indicated by the list of dead: see, *Monmouthshire Merlin*, 22 February 1862.
127 *Cardiff Times*, 21 February 1862.
128 *Merthyr Telegraph and General Advertiser for the Iron Districts of South Wales*, 23 December 1865.
129 Church, *British Coal Industry*, 275.
130 Rymer, *Martyrdom of the Mine*, 6.
131 Ibid.
132 Colls, *Pitmen*, 25–30; Campbell, *Lanarkshire Miners*, 26–45.
133 Gleeson, *Geographies of Disability*, 106–7; E. P. Thompson, 'Time, Work-Discipline and Industrial Capitalism', *Past and Present*, 38 (1967), 56–97.
134 PP 1842 (382), 653.
135 Morris and Williams, *South Wales Coal Industry*, 231–2.
136 *The Examiner*, 18 August 1833. The same report also appeared in *The Bristol Mercury*, 24 August 1833.
137 PP 1842 (380), 111.
138 Church, *British Coal Industry*, 238, 251.
139 Campbell and Reid, 'The Independent Collier in Scotland', 69.
140 Daunton, 'Down the Pit'; Morris and Williams, *South Wales Coal Industry*, 57–62; Flinn, *British Coal Industry*, 82–91; Duncan, *The Mineworkers*, 72–7.
141 Church, *British Coal Industry*, 329–36; Martin Daunton, 'Down the Pit: Work in the Great Northern and South Wales Coalfields, 1870–1914', *Economic History Review*, 34:4 (1981), 582.
142 James Thain, 'The Memoirs of James Thain of Stepaside, Kilgetty (born 3 May 1870)' in M. R. Connop-Price, *Pembrokeshire: the Forgotten Coalfield* (Ashbourne: Landmark Publishing, 2004), 224.
143 Daunton, 'Down the Pit', 583–4.
144 David Tonks, 'A Kind of Life Insurance: the Coal-Miners of North-East England 1860–1920', *Family and Community History*, 2:1 (1999), 45–58, especially 49.
145 Ben Curtis and Steven Thompson, '"This is the country of premature old men:" Ageing and Aged Miners in the South Wales Coalfield, c. 1880–1947', *Cultural and Social History*, 12:4 (2015), 587–606.
146 Colls, *Pitmen*, 29–32.
147 Rymer, *Martyrdom of the Mine*, 6.

2

MEDICINE AND THE MINER'S BODY

There were two prevalent views about mineworkers' bodies in Victorian Britain. On the one hand, miners were represented as a distinctive class of workmen, prey to numerous diseases 'induced by the very unwholesome nature of [their] occupation'.[1] On the other, coalmining was represented as healthy work and miners were admired for their physical robustness.[2] These contrasting ideas about the health effects of coalmining shaped public perceptions of the industry as well as miners' view of themselves. As a contributor to the *Miner and Workmen's Advocate* wrote in 1863, miners were 'strong and active men', but 'pallid in complexion and bent in form, by reason of excessive labour, heat, and foul air, which they are constantly obliged to breathe'.[3] From the late eighteenth century onwards, the physical characteristics of coalminers and the particular health problems associated with underground work, came under increasing scrutiny. The 'habits and diseases' of miners were increasingly captured in a net of professional narratives by doctors, policymakers and social reformers which revealed the ways in which the health and occupational illnesses of colliers compared with those working in other sectors of the industrial economy. This work drew attention to the manifold causes of illness and incapacity in mine work beyond the accidents that prompted government inspection, suggesting a much wider experience of disablement in the coal industry.

This chapter charts and explains this growing interest in the bodies of mineworkers, placing it in the context of broader campaigns for public health and industrial reform. Focusing in particular on the services provided through workplace 'sick clubs', the chapter examines the development of medical responses to sickness and injury in and around coalmining communities in late eighteenth- and nineteenth-century Britain and shows how the coal industry was innovative both in the extent of medical provision available to workers and in a variety of responses to workplace injury from first aid to specialist convalescent homes. The expansion of medical services made mineworkers, like other industrial workers, increasingly subject to medical surveillance. Yet the history of medical intervention in the lives of Britain's nineteenth-century coalminers is not simply one of increasing professional authority over the bodies of industrial workers. Like other working people, sick and disabled miners and their families might demonstrate agency and independence in their dealings with doctors.

Mining bodies

The different working conditions within coalfields and between them made the overall physical health and appearance of miners difficult to generalise. When parliamentary commissioners gathered information on a pit-by-pit basis as part of the great inquiry into the employment of children in the early 1840s they discovered a good deal of variation in the incidence of disease, deformity and perceptions of good or ill health between mines. Mr Atkinson, a surgeon employed by Wylam Colliery in Northumberland, reported that 'pitmen have more sickness here perhaps than at many other collieries' on account of its low position near the river Tyne and the smoke coming from the adjacent ironworks and coke ovens.[4] As we saw in the previous chapter, conditions underground varied considerably and factors such as the thickness of coal seams were deemed to have a decisive effect on the physical characteristics of those who worked them. Those men who worked in seams of 'sufficient thickness to permit the free use of muscular action' were 'erect and of good figures', but those cutting coal in narrower seams 'have the spine permanently curved, and the legs frequently bowed'.[5] The variety of risks associated with working in different seams, the differing qualities of the coal mined and the varying degrees of exertion required to extract it meant, as J. T. Arlidge wrote in 1892, that 'data respecting the health of colliers for one district, or even one pit, compared with another, are not of general application'.[6]

The variation in conditions was reflected in differing experiences of occupational mortality in nineteenth-century mining. As Anthony S. Wohl has noted, whereas the death rate among men aged 45–55 among the south Wales miners at the end of the nineteenth century was 24.47 per 1000 living compared with a national average of 21.37, in north-east England it was only 16.35.[7] Setting aside the high mortality in mining caused by accidents, experts on occupational diseases such as Arlidge regarded conditions in coalmining as relatively favourable to health. Miners, he wrote, 'escape the evils of sedentary work; their hours of labour are shorter than those of many indoor occupations' and the 'circumstances of their employment is a bar to riotous living'. Working underground preserved miners from inclement weather, providing a more 'equable climate than outdoor workers'.[8] Many miners received allowances of free or cheap coal from their employers to heat their homes and these were also seen as contributing to the 'healthy' environment in which colliers and their families lived.[9] The dirty nature of underground labour furthermore made it necessary for miners to adopt regimes of hygiene and cleanliness, which were seen as improving their overall health. Although opinions on miners' hygiene varied, evidence supplied by medical officers of the Merthyr Tydfil Poor Law Union for Edwin Chadwick's *Sanitary Report* of 1842 suggested that the 'health of the colliery population was very good – a circumstance which is ascribed to their habitual cleanliness'. Washing in a tub on return from work made miners less liable to 'cutaneous disease' than other workmen who 'do not wash so completely or thoroughly'.[10]

Above all, the arduous nature of mine work, from hewing coal to hauling it through underground passages and winching it to the surface, contributed to the idea that those who entered the pits were selected on the basis of their physical strength, making them more resilient to disease or constitutional weakness than others. James Essex, a surgeon from Pontypool echoed the view of many other medical witnesses to the 1842 Children's Employment Commission when he stated that coalminers were 'superior' in 'physical strength' to those employed in agriculture. They were also 'capable of enduring more fatigue', since labour was apportioned to a worker's strength.[11]

By the end of the nineteenth century, the idea of colliers as 'picked men physically' was commonplace. 'As the work of a collier requires considerable bodily vigour' explained Arlidge, 'it will be taken up by the stronger members of a community'.[12] However, although 'extraordinary muscular development' was regarded as one of several physical characteristics that distinguished miners from the general population, miners' 'muscularity' was not necessarily regarded as a source of health or general able-bodiedness.[13] In his pioneering study, *The Effects of the Principal Arts, Trades and Professions … on Health and Longevity* (1831), Charles Turner Thackrah described ways in which hard physical work done by coalminers around his native Leeds caused 'deformity' since 'one set of muscles is immoderately and almost constantly exerted, while another wastes for want of action', causing other defects such as spinal curvature.[14]

As the coal industry expanded, the physical distinctiveness of miners became a matter of public discussion. The human scenery of a 'northern coal district' was, according to a *Penny Magazine* article of 1835, as notable as its physical geography. Pitmen, 'black as sweeps', were 'in a great measure

[set] apart from other classes of the community', a 'distinct race from the neighbouring peasantry'. For all miners, working in darkness meant that their complexions, when visible beneath the grime, were 'generally sallow and unhealthy', their eyelids often 'swollen' and their eyes 'diminutive' in appearance. The article contended that the 'physiognomy of miners' was not 'of a very intellectual cast' and was distinguished by high cheekbones, 'great width of the middle part of the face, and an angular form of its lower portions'.[15] The description of miners as a specific 'race of men' grew in popularity during the nineteenth century, reflecting a growing tendency to elide discourses of race and class in commentaries on British workers in general and anxieties about the ways in which underground work brutalised miners and permanently altered their bodies in particular.[16] 'Small bulk of body, paleness and angularity of visage, and their general appearance, which is far from robust, would lead to the conviction that they are a somewhat deteriorated race,' remarked John Roby Leifchild in his report presented to the Children's Employment Commission of 1842. Living in close-knit communities with a tendency to intermarriage meant that 'natural and accidental defects' were passed on by inheritance to miners' offspring.[17] Disease and 'deformity' were therefore not simply a matter of working conditions; they were increasingly viewed as having a hereditary basis.

The distinct physiological characteristics of miners' bodies were also inextricably linked to their mental attributes. 'The nervous system, including various parts of the brain,' wrote Dr S. Scott Alison in his remarks on Scottish miners, 'are comparatively little exercised, while that of the muscles is inordinately overworked.' Consequently, 'the collier becomes more a mining or working animal than a thinking being – more a machine than a rational creature'.[18] Mine work, therefore, not only had a detrimental effect on the physical health of miners, it also degraded their mental faculties and sensitivity and effectively turned them into brutes. 'When we consider the mode in which hour after hour of the miner's gloom and monotonous existence is spent, in darkness ... frequently pursuing their working in silence and solitude,' wrote G. Mallett in the *Association Medical Journal* in 1855, the effect on the miners' physical and mental faculties was bound to be damaging:

> A life so spent must exhibit few opportunities of calling out and cultivating the sensibilities of the nervous system; on the contrary, the tendency must be to depress its natural activity, and to render it less sensitive, in short, apathetic.[19]

The effect, according to this author, was not only that the majority of miners were 'very low in the scale of intellectual culture', but also that miners lacked nervous sensibility to pain. Miners, argued Mallett, were capable of recovering from 'such an amount of injury as would in all human probability have proved fatal if it had occurred to individuals *occupying a higher social position*', or to those whose 'intellects [were] more highly developed'. Among the examples he cited to support this argument was the case of John Isherwood whose right leg had to be amputated after being run over by a railway coal wagon – an operation he bore 'without any expression of pain'. The hierarchical nature of sentience, what Joanna Bourke describes as the 'great chain of feeling', was a powerful idea in nineteenth-century Britain. Differing sensitivity to pain distinguished humans from animals, and more 'civilised' human beings from others. Exposure to the 'brutish' nature of hard manual labour distinguished the feelings of the working man from those of his social 'superiors'. Insensibility to pain also had a marked racial element, differentiating 'savages' from 'cultured' Europeans. Significantly, Mallett likened colliers to North American 'Red' Indians who were also known for their 'calmness' in times of physical hardship on account of their 'nervous systems being less developed'. By doing so, he implied that miners similarly constituted a 'race' delineated by their special physiological characteristics whose capacity for endurance in the face of the traumatic and disabling effects of their work denoted their bodily 'otherness'.[20]

Miners' diseases and the production of medical knowledge

Professional interest in the effects of mine work on the minds and bodies of those employed in it grew as the industry expanded during the nineteenth century. However, European men of science and letters had long identified miners as a class of workmen susceptible to special kinds of occupational disease and injury – a theme that can be traced from antiquity through the pioneering work on the diseases of mineral miners written by Paracelsus and Agricola in the sixteenth century.[21] Between 1600 and 1800 more than twenty-five authors wrote observations on mining diseases, most notably Bernardino Ramazzini, who devoted an entire chapter to miners' diseases in his *De Morbis Artificium Diatriba* (1700).[22] Ramazzini's theory that the position and motion of workers' bodies was a principal cause of occupational illness or deformity was especially influential on assessments of the deleterious effects of industrial work in the eighteenth and early nineteenth centuries.[23] In relation to mining, Thackrah argued that the tendency of coal hewers to spend a long time in a bent sitting position, damaged circulation, causing ill health and impairment.[24] Several medical witnesses to the 1842 Children's Employment Commission likewise attributed some degree of deformity in coalminers to their working posture and often linked this to other medical conditions. For example, T. M.

Greenhow, surgeon of Walker Colliery, Newcastle-upon-Tyne, noted that colliers had a 'bent and cramped character' of body caused by their working environment, while others noted that miners' 'prevalent deformity', a curved spine caused by their working position, could lead to chronic diseases of the stomach and liver.[25] Later on renal diseases were also associated with the 'doubled up' posture in which those cutting coal worked.[26] By linking physical deformity to susceptibility to illness in such ways, the postural theories informing nineteenth-century discussions of miners' health blurred the distinction between 'disease' and physical impairment.[27]

Early works on occupational health may have focused on the diseases associated with mineral mining, but, by the end of the eighteenth century, *coal*mining's increasing economic importance meant that studies of miners' diseases began to concentrate more fully on the health of colliers specifically.[28] The first full-length study devoted to the injuries of colliers was Edward Kentish's *Essay on Burns* (1797), which inquired into the best means of treating coalminers burnt in the mining explosions that were becoming more common in the deeper mines of north-east England at this time.[29] Kentish's work addressed what he saw as neglect by the medical faculty of coal workers that stemmed from their feelings of 'disgust' about mining as a dirty and unpleasant occupation, hidden from sight underground.[30] He addressed his work to the 'Proprietors of Collieries Upon the River Tyne', whom he regarded as the 'natural guardians of the health and comforts' of their workers, situating the medical treatment of miners' injuries within a nexus of paternalistic duty of care.[31] Recognition of coal as an industry 'of the first-rate importance' also motivated Robert Bald's study of the Scottish coal trade published in 1812, which included a section on the medical consequences of women's involvement in underground work as coal bearers, an activity he saw as 'prejudicial to their health' and to that of their neglected infant children.[32] By the 1830s and 1840s the growing employment of medical professionals by coalmining companies, discussed in more detail later, had produced new experts in the diseases and disabilities of coalminers. These medical professionals were given a voice through their contributions to government inquiries such as Chadwick's Sanitary Commission and the Children's Employment Commission, both of which reported in 1842.

The ways in which medical men's increasingly close involvement in the lives of diseased and disabled miners stimulated the production of medical knowledge is evident above all in the expanding nineteenth-century literature on lung diseases. While the influence of dust in causing respiratory illness had been noted by writers in the sixteenth century, it was during the 1820s and 1830s that the influence of working conditions on the incidence of 'miner's asthma' or 'black lung' came under increasing scrutiny. No doubt the interest of the medical profession in the respiratory illnesses of miners was piqued by the expansion of the industry and the growth of deep mining, which increased occupational morbidity. But the fact that the notoriously dry and dusty Scottish coalfield was adjacent to the universities of Edinburgh and Glasgow, two of nineteenth-century Britain's most influential centres of medical learning, also helped ensure that sick and disabled miners became compelling objects of medical investigation.[33]

Historians of medicine have documented the ways in which pioneers such as James Gregory, Matthew Gibson, William Thomson and George Steele sought to extend the practice of pathological anatomy developed in the medical schools of Paris at the turn of the century to the investigation of conditions such as melanosis in the lungs. But while these accounts have presented these pathological advances in terms of medical progress, what is interesting from a disability perspective is the ways in which medical research increasingly targeted the chronically ill coalminer as a source of knowledge.[34] In his 1837 article, 'On Black Expectoration, and the Deposition of Black Matter in the Lungs, Particularly as Occurring in Coal Miners etc', that reviewed the medical progress of the previous decade, Dr William Thomson of the Royal College of Physicians and Surgeons in Edinburgh described how he and his father had spent many years studying local coalminers and iron moulders and initiated calls to 'professional gentlemen in different parts of the country' to supply evidence based on their own treatment and observations of colliers.[35]

Thomson's account documented vividly the progressively disabling effects of lung disease. For example, the case of George Hogg, a miner at Collinshiel Colliery near Bathgate, was brought to Thomson's attention by Dr James Y. Simpson. Despite being 'tall and of a very athletic form', Hogg had become 'unable to follow any active employment' due to problems with his breathing which, 'even when he is at rest is somewhat laborious and sonorous'. He was able to undertake 'some slight work in his garden' but this led to fits of coughing in which he would bring up '2 or 3 profuse sputa' the colour of 'black ink'. During one such attack prior to his death he had spat up a 'Scotch mutchkin, or nearly fifteen fluid ounces' of mucus. Thomson did not mention what treatment, if any, Simpson was able to give Hogg, but noted with interest that upon receiving news of the collier's death Simpson had 'obtained permission of the relatives to inspect the chest'.[36] Cases such as this, where the observed progressive degenerative symptoms of lung disease could be followed with dissection, were of most value to the advancement of medical knowledge. In 1835 Simpson had been able to send part of the lungs of another miner,

Robert Leishman, which exhibited black membrane, to the Museum of Guy's Hospital in London as a significant medical specimen.[37]

Yet Thomson's report also revealed the difficulties faced by medical men in obtaining specimens caused by the resistance of families of the deceased. Those who in their own estimation stood for medical progress in the identification of the causes of lung disease found themselves opposed by what they described as the 'deep-rooted prejudices against anatomical examinations entertained by the coal-miners'. As one Dr Dewar wrote in a letter to Thomson about diseased miners at Hallbeath Colliery in Fifeshire, 'To dissect a collier is *periculosae plenum opus aleae*' (a work full of dangerous risks).[38] Dissection's historical association with the punishment of criminals, combined with recent memories of scandals such as the Burke and Hare murders of 1828 that had provided cadavers for the Edinburgh Medical School, made dissection unpopular and cast suspicion on the motives of anatomists. Set against the backdrop of the 1832 Anatomy Act, which increased the supply of bodies for dissection from workhouses and other institutions and bolstered the power of medical men over the bodies of the poor, pathological investigation held potential to bring doctors into conflict with families.[39] While doctors dismissed the opposition of some mining communities to having the bodies of the diseased anatomised as deriving from their 'prejudice', this resistance might conversely indicate respect for the dead, love for disabled family members, refusal to see them simply as objects of medical curiosity and a desire to protect their bodily integrity.

The growing attention to respiratory illnesses in miners may also be seen as symptomatic of a shift away from humoralism that explained illness or injury in terms of 'vitiated constitutions' to new modes of thinking that located the origins of disease and deformity in specific organs of the body.[40] Although medical witnesses still relied heavily on general notions of workers' constitutions when discussing the health effects of factory work during the intense debates over industrial reform in the 1830s, by the time of the publication of the 1842 Children's Employment Commission the focus was shifting towards localised medical conditions.[41] The commissioners amassed significant medical evidence on the effects of mining on workers' bodies, documenting alongside the manifold accidents that robbed miners of their eyes and limbs many diseases that had a disabling effect on coal workers. 'Bad breath' – a term that captured colloquially a number of lung diseases – was frequently reported among Scottish miners.[42] Inflammatory diseases of the lungs and rheumatic fever were deemed common, and attributed not just to dust, but also to perspiration.[43] Some witnesses, such as R. P. Edger the salaried surgeon at Hetton Colliery in County Durham, believed pre-existing chronic conditions such as asthma to be exacerbated by underground working, although much depended on the quality of ventilation in particular mines.[44] 'Bad air' in collieries was also linked to a variety of problems including sickness, vomiting, headaches and breathlessness.[45] Working in heated conditions could lead to painful boils, attributed to excessive consumption of water.[46] Some medical witnesses pointed to the increased risk of hernia caused by lifting heavy weights.[47]

The report provided much evidence of the harmful effects of mine work on children's physical and mental development. It was claimed that child trappers working in the dark 'become almost idiotic from the long, dark, solitary confinement', while many others suffered from fatigue, aching bodies and nausea. Physical exertion produced numerous complaints harmful to growing bodies, from the risk of rupture to heart disease.[48] All suggested, as S. Scott Alison reported to the commission, that the 'physical condition of the boys and girls engaged in the collieries is much inferior to that of children of the same age engaged in farming operations, in most other trades or who remain at home unemployed'.[49]

Witnesses also testified to the damage of working underground on women's health. In Scotland, many reported miscarriages and premature labours caused by the arduous tasks performed by women hauling coal, and some articulated a distinctly female concept of disability related to underground labour based on women's inability to bear healthy children. Jane Peacock Watson told east Scotland Sub-Commissioner Robert Franks that during the thirty-three years she had worked underground two of her nine children had been 'dead born', which she attributed to the 'oppressive work'. She claimed that a 'vast [number] of women have dead children and false births', for it was common for women to 'work below till forced to go home to bear the bairn'. Minework, she said, 'ruins the women; it crushes their haunches, bends their ankles and makes them old women at 40'.[50]

However, not all medical witnesses were equally convinced of the dangers of underground work to women's reproductive health. William Brownlee, surgeon to Shotts Pits and Collieries in the west of Scotland, recalled the case of a 'young married woman who had a premature birth from an accident, and was some time out of health from it'. However, he did not think mine work 'injurious to [women's] health'.[51] Furthermore, as Frank Jowin, surgeon to the Ebbw Vale Iron-Works Company in south Wales, reported, the work given to women at the mines in his region was not likely to 'produce distortion of spine or deformity of pelvis', and out of an estimated '1400 or 1500' childbirths he had been asked to perform he had 'never been called upon to deliver a woman with instruments whose labour was retarded by a deformed pelvis'.[52]

Indeed, despite the conclusion drawn by one critic of the conditions in

Britain's collieries that 'the evidence collected in almost all the districts proves too often that the collier is a disabled man' by nature of his exposure to chronic disease, accident and the progressive deformity brought on by hard labour, the medical evidence presented to the 1842 Commission was often contradictory, reflecting fully the opposing views of miners' bodies as both healthy and diseased prevalent in nineteenth-century Britain.[53] For example, for all those who highlighted the dangers of dust and poor ventilation to the health of miners' lungs, there were others such as George Eliot, the head viewer at Monkwearmouth Colliery in South Shields, who regarded his colliery as 'quite an asylum for asthmatic people; and an asthmatic man who cannot possibly work at bank can work well below', due to the constant temperature in this exceptionally deep mine.[54]

While the 1842 Commission's report provided the most extensive account of mining disease and disability published to date and presaged the system of government inspection of safety in mines to reduce *accidents*, policymakers remained uninterested in the fatal or disabling effects of *diseases* – a point recognised by J. B. Thomson in a paper published in the *Edinburgh Medical Journal* in 1858. The evidence of official reports 'setting forth that miners were short-lived, and subject to frequent and fatal maladies peculiar to their calling', had not led to much in the way of greater 'protection and sympathy' for this 'long neglected class'. It was, he argued, in accord with the 'principles of a sound economy, and the dictates of our common humanity' that a medical inspectorate should be established to work with mining engineers to 'apply the most enlightened rules of hygiene for the safety and health of this numerous and important class of work-people'.[55] Thomson was a rare voice in calling for official intervention to improve the health and well-being of colliers in ways that tackled their propensity to chronic disease *in addition to* their susceptibility to accidents.[56] However, if policymakers were unwilling to introduce medical surveillance of miners, at a local level colliers came into contact with medical services in a variety of ways. The remainder of this chapter examines relationships between doctors and coalminers within coalfield communities and asks what medical treatments were available to those who worked in the coal industry.

Accessing medical services

Like other workers in industrialising Britain, mineworkers accessed healthcare within a mixed economy of medicine, which included timeworn family remedies and unorthodox healers as well as contact with medical professionals provided by both the state (via the Poor Law) and voluntary agencies. Self-help organisations such as friendly societies were part of this mixed economy of care. Even though most were initially concerned primarily with workers' financial well-being, many friendly societies also offered medical attendance to their members and this function grew as the nineteenth century progressed. Perhaps the most striking feature of the mixed economy of care, however, was the provision of medical assistance for coalminers via workplace schemes.[57] Mineworkers were one of the first sections of the British working class to become accustomed to the services of physicians and surgeons. Coal owners were major employers of doctors at a time when the medical profession was still trying to establish itself and this helps explain why Kentish dedicated his treatise on burns to them. Fifty years later, Mines Inspector Herbert Mackworth told an audience at the Bristol Mining School that it was axiomatic that colliery managers should employ specialised surgical expertise, informed by an understanding of 'the proper treatment of the diseases to which colliers are liable'.[58] Within the coal communities of Britain, doctors built up close relationships with patients. These, however, were not always harmonious, as we shall see.[59]

The establishment of specialist medical practice associated with collieries developed in an ad hoc manner as the coal industry expanded over the late eighteenth and nineteenth centuries. Kentish's work as a surgeon at coal mines in the late eighteenth-century Tyne and Wear district is an early example of the employment of medical practitioners as the industry expanded. Similarly, in parts of the Rhondda Valley in Wales medical practitioners were attached to collieries soon after coal was discovered there in 1809.[60] Evidence presented to the Children's Employment Commission in the early 1840s provides the most complete overview of how medical services had evolved by the middle of our period. Thomas Alexander Cockin, manager of a colliery called Pease's Deanery or Adelaide Wallsend in the Auckland area of Durham, explained that the mining company employed a surgeon 'in case of accident' who was contracted at a salary to the mine.[61] Some medical men were employed by a variety of collieries at the same time – the Newcastle surgeon Mr Heath reported that he was employed by four collieries in that district.[62] Sometimes, colliery doctors also served the wider community as Poor Law Medical Officers or combined their colliery appointments with work for other industries or railway companies.[63]

Salaried surgeons were employed primarily to deal with accidents, for which the expenses were 'defrayed by the [mine] owners'.[64] Other medical complaints were dealt with by doctors funded out of miners' wages through workplace sick clubs. As the surgeon of Monkwearmouth Colliery, W. J. Dodd, explained, he was employed 'by the owners for colliery accidents' but he also 'attends in ordinary cases of sickness on the principle and payment of the sick fund prevailing through the colliery districts', for which men had sixpence per fortnight deducted from their wages.[65] Workplace sick clubs were established under the 1831 Truck Act. This 'gave certain employers the right to provide medical attendance and medicine for their employees and empowered them to make deductions' from workers' wages to do so, provided their employees consented.[66] While some mine owners may have pressured miners to join clubs, others seem to have allowed workers a truly free choice in the matter. As Dodd explained, the fortnightly medical deduction was 'quite an optional arrangement on the part of the men'. He blamed the improvidence of miners for necessitating such a 'plan'.[67] Furthermore, the sense of 'ownership' of the medical attendant's services might be seen differently from pit to pit. For example, at Willington (County Durham), a surgeon 'resides at the colliery, who is employed by the men exclusively'. Miners contributed towards his services at the rate of about sixpence a week for a family, or four pence for single men, giving him an annual income of around £80. In contrast 'the proprietors employ a surgeon for all accidents, who resides at Newcastle, and is paid by his visits at accidents'.[68]

In south Wales, miners paid for medical care via the 'poundage' system whereby employers took a levy from every pound earned to pay towards a surgeon.[69] At Loughor Colliery, which employed 50 people at the time of the Children's Employment Commission, there was a surgeon 'appointed to attend both the men and their families in cases of accidents or sickness'. The surgeon's contract included wives and children as well as the men who worked at the pit, but excluded midwifery cases, and was funded by payments of sixpence a month levied on the wages of men and boys alike, except for the 'door-boys' who paid three pence each.[70] While sick clubs were established in response to high levels of injury and disease, some struggled to cope in times of high demand. At Risca Colliery in Monmouthshire, so many men were in receipt of medical assistance that the club was in debt to the employer who underwrote the scheme.[71] In parts of south Wales, workplace 'sick clubs' did not merely provide access to medical services. In Merthyr Tydfil, for example, the sick fund paid for by workers at the Dowlais Company contributed to 'paying the surgeon who attends the men, for supporting the sick workmen, and for paying the schoolmaster'. Indeed, the fact that the school at Dowlais was attended by 'about 12 boys maimed and crippled', otherwise 'incapable of labour', shows how funded medical care might go beyond the injured or sick body itself and extend to welfare and education, helping to fund services that would ultimately aid the rehabilitation of those 'crippled' in accidents by helping them find alternative employment.[72]

By the middle of the nineteenth century, miners across Britain were accustomed to paying into workplace schemes that provided access to a surgeon as well as (in some cases) funding other services such as schooling or the provision of a reading room.[73] Such provision was part of the reciprocity between workers and employers in coalmining and demonstrated a strong paternalistic ethos. The 'exceptional' services enjoyed by colliers through workplace 'sick clubs' and the 'generosity' of colliery owners was praised in the press. 'The Collier at Home', an article published in *Household Words* in 1857 supposedly written by a surgeon, presented an idealised view of miners' medical care in which they could always trust in the 'liberality' of the company sick fund to care for their needs and rely not just on the attention of doctors, but also on the 'sympathy' of their employers:

> I have never seen anywhere so distinctly as among the mines, the rich helping the poor, knowing them all personally, visiting them when sick, and sorry without ostentation or intrusion – looked upon them as helpers and friends without any mean or cringing flattery.[74]

Nevertheless, the expansion of sick clubs also demonstrated the power of medicine as a tool of workplace discipline, extending the employers' control over their employees both by obliging them to subscribe to a compulsory fund and in determining their entitlement to care.

Medical treatment

If the employment of surgeons by coal owners to treat bodies burnt, crushed, dismembered or lamed in the mines shows a recognition both of the dangers of mining and of a paternalistic duty of care, medical responses to accidents developed in a haphazard manner during the nineteenth century. Special rescue equipment or teams of trained personnel (with the exception of surgeons) did not appear as a regular feature on the mining landscape until the closing decades of the century. Prior to that time, victims of accidents were often brought up to the surface by their workmates, using rudimentary apparatus such as coal corves or baskets to transport the wounded.[75] Mines Inspector Herbert Mackworth delivered a scathing verdict on the emergency facilities and procedures at the mines he had visited in a lecture given to Bristol Mining School in 1857. He called on mine owners to ensure that medical equipment and supplies were 'always on hand', particularly 'restoratives' and 'properly constructed litters and bandages'. He was especially concerned for the comfort of injured mineworkers. 'Too often', he claimed, may the rude, jolting, clumsy cart be seen wending its way through a mining village, containing some unfortunate miner with broken limbs, on a bed of rough straw or of the work clothes of some of his more humane fellow-workmen, every stiff jarring motion of the cart producing fresh agonies to the sufferer.[76]

In Mackworth's opinion, moreover, such scenes were not isolated incidents, but the norm, 'allowed and followed in nearly every colliery district in the country'.[77]

Before the 1870s, it was rare to find ambulance services or equipment at collieries. Rescuers improvised with what was at hand. Carts, wagons or doors as makeshift stretchers were often used to convey wounded miners to their homes or hospitals. At around the same time Mackworth made his comments in Bristol, others were also highlighting the suffering of miners hurt in accidents and calling for more comfortable and dignified means of conveying the wounded from pithead to sickbed. For example, in June 1858 the *Colliery Guardian* endorsed the recommendation of another mines inspector, Matthias Dunn, for 'spring palanquins for the conveyance of wounded men either to their own homes or to an infirmary'. By reducing jolting, these 'palanquins' or hammocks, carried at shoulder height by four men, would minimise the pain of injured miners and were preferable to carts, which were little more than 'instrument[s] of abominable torture' when used to transport a 'man with a broken limb or scorched skin'. The pain suffered in this way, and risk of further damage to the injured body that might hinder full recovery or lead to permanent disability, provided compelling reasons for the adoption of such methods of conveyance.[78]

However, the issues went beyond medical efficacy or patient care. The transportation of the injured or dead from the pithead was a pivotal scene in the emotional drama of nineteenth-century mining accidents.[79] They were public spectacles that simultaneously displayed the horrors of mining and the self-sacrifice of miners. Provision of palanquins would, Dunn argued, demonstrate 'forethought and sympathy' on the part of coal owners and would 'do much towards establishing a friendly feeling between the employers and their servants'. They would have a 'happy effect on public opinion' since there were 'few things more revolting than the sight of a clumsy cart jolting through the streets, which is known to contain the mangled remains or the suffering body of some poor collier'. Effective emergency provision was essential to the dignity of the miner as well as the preservation of his life or limbs.[80]

Mines inspectors' calls for better equipment in emergency care were eventually taken up by the St John Ambulance movement in the 1870s, which made suggestions for improved 'litters' to carry injured miners based on those used on the battlefield. That military methods of conveying the wounded were thought appropriate for use in colliery accidents indicates both the practical and rhetorical links between warfare and mining in the popular imagination of Victorian Britain. Given the considerable dangers of mine work it was an easy and apt association to make and one the ambulance movement was happy to exploit to further its goals. As war promoted innovations in emergency care and equipment, mining provided a suitably hazardous environment for the continued testing, development and application of new, military inspired, medical technologies and methods in the civilian world. Given this, mining in nineteenth-century Britain deserves to be regarded, as it was at the time, as an important bridge between military and industrial health and safety regimes. By facilitating the transfer of approaches to injury derived in wartime to the industrial workplace, mining helped shape civilian responses to accidents in peacetime.[81]

As calls for the provision of stretchers and other equipment at mines increased during the nineteenth century, so too did appeals for first-aid training for colliery staff. The value of practical knowledge of first aid and accident management had been espoused by eighteenth-century physicians such as William Buchan and Samuel Tissot, and by the Humane Society, established in 1774.[82] From this period, surgeons attending mining accidents started to become renowned for their expertise in providing emergency responses to trauma.[83] For example, after an underground fire at a colliery in Llansamlet near Swansea in 1787, two surgeons – one of them 'a pupil of the benevolent founder of the Humane Society' – managed to revive eight mineworkers who were brought to the pithead presumed dead.[84] By the mid-Victorian period the heroic exploits of some colliery doctors attending to the victims of disasters earned national admiration. 'Colliery surgeon' Dr Davidson 'scarcely left the pit's mouth night or day for the first four days' after a catastrophic accident at Hartley Colliery in 1862 left many men trapped underground. Despite their best efforts, however, it was rare for medical men to arrive immediately.[85] Mackworth drew attention to this situation in his Bristol Mining School lecture and urged colliery managers to study 'the diseases and accidents to which miners are subject, and the best mode of treating them until professional medical assistance can be obtained'.[86] Before the campaigns of the St John's Ambulance movement in the 1870s, which, alongside its calls for better emergency equipment at mines, called for first-aid training for miners, few mineworkers received any formal instruction in how to treat injured colleagues.[87] Throughout the nineteenth century, then, the quality and effectiveness of the emergency treatment injured miners received from first responders, who were usually their workmates, was highly varied and largely a matter of luck.

Diseased and injured miners and their families drew on a patchwork of care incorporating elements of both formal and informal medicine. In the late eighteenth century, lay healers skilled in humoral therapies such as purging and bloodletting were an important part of the medical landscape of coal communities. In his autobiography, Anthony Errington remembered his schoolmistress as a 'good Doctriss, scield [sic] in Leting Bleed'.[88] The expansion of formal medical provision in the coalfields during the nineteenth century did not fully displace this reliance on unorthodox healers. On the one hand, some miners embraced their access to medical professionals via sick clubs enthusiastically, anxious not to see their subscriptions go to waste. As Edward Robatham, surgeon of Risca in south Wales, told the 1842 Children's Employment Commission, since the colliers 'have a doctor to apply to in every instance of necessity, they are also in the constant habit of taking aperient medicines, whether they require it or not, imagining that they must have something for the money they monthly pay to the doctor'. Robatham himself was in the 'habit of supplying [medicines] freely, feeling assured it has a tendency to ward off disease'.[89] On the other hand, colliery doctors and others frequently complained of miners' enduring 'superstition' in medical matters. William Morison, who provided medical services to the Countess of Durham's collieries in the 1840s, described the pitmen of north-east England as 'persons whose minds are singularly warped by prejudices'. He argued that in coalfield areas doctors spent more time trying to 'ward off the pestiferous influence of old women's nostrums and crochets' than tackling diseases themselves. He cited the example of a 'medical gentleman in the county of Durham' who attended a boy wounded by a pick. The boy's family kept the bloodied implement next to his bed in order to see whether the blood on the point would rust – an apparent sign that 'the wound in that boy's back will canker and he will die'.[90]

Pits situated near to larger towns such as Newcastle were better served by formally trained medical practitioners than those in more remote areas where bonesetters, charmers and irregular healers were quick to move in where doctors were thin on the ground.[91] Attempts to stamp out irregular practice, such as the 1858 Medical Act, which required registration of medical professionals, did little to deter unorthodox medicine in mining areas.[92] Miner Edmund Stonelake described the situation in south Wales at the turn of the twentieth century, where 'every village and town' was visited by confidence tricksters claiming to be able to 'set bones, draw teeth, remove corns and bunions, cure deafness, rheumatism and almost every complaint that human flesh is heir to'.[93] With the development of publications aimed at coal workers such as *The Miner and Workmen's Advocate* came opportunities to advertise medicines and self-help guides offering cures for perceived common afflictions of colliers, including those which caused 'debility' or impairment. The issue for Saturday 13 June 1863, for example, contained advertisements for 'Grimstone's Celebrated Eye Snuff' and for various products aimed at reducing 'debility' and the 'premature decline of man' – the result in this case not of disabling mine work in particular, but rather the more universal 'secret sins of youth'.[94] Thus, as interest in the peculiar 'habits and diseases' of mineworkers grew, stimulating new medical knowledge and practices, so too did the realisation that miners constituted a valuable market for the services and products medical practitioners and innovators were offering.

There were clearly quite serious limits to what orthodox and folk medicine throughout the period could realistically achieve despite the claims of optimistic healers. Beyond minor surgical procedures, dressing and cleaning wounds and palliative care, expectations that medical practitioners could effect cures, in the modern sense of the word, were low. Indeed, the recurrent cases documented in official enquiries of mineworkers returning to work when 'lamed' may be seen not just as evidence of people 'working through' their impairments or ill health, as discussed in the previous chapter, but also as recognition that 'curing' someone was more about relieving their symptoms sufficiently to enable them to return to work than restoring them to full function.[95] Even the management of pain was challenging. Until the emergence of modern analgesics, laudanum and morphine were the main means of controlling pain. Kentish documented the widespread use of opiates to relieve pain in mining communities during the late eighteenth century.[96] How frequently mining families could access, or indeed afford, these drugs, however, is difficult to ascertain. It seems likely, though, that few working-class Britons who experienced chronic pain would have enjoyed a ready and uninterrupted supply of opiates. The belief, then, that miners were impervious to pain may have proved a serviceable myth in an era when pain management was basic.[97]

The physical trauma of accidents and injuries was easy for nineteenth-century Britons to perceive. Although the limits of medicine were clear, those treating ill health and injury in the coal industry had an idea about how to respond to physical trauma and an expectation of some success. When it came to the psychological consequences of mining accidents, however, coalfield communities seem to have been less certain about what to do. Among several survivors of the infamous Wallsend disaster, which killed more than 100 men and boys in 1835, at least two were described by eye-witness James Everett as 'delirious' and 'incoherent'. While Everett's account gives details of the treatment of survivors' physical injuries, there is no mention of how these two mineworkers' presumably psychological traumas were addressed.[98] Such silence suggests no medical intervention was offered to help heal survivors' apparent mental scars.

Locations of care

There were two main sites of treatment for sick or injured mineworkers in the nineteenth century: the home and hospitals. Of the two, the home was undoubtedly the most significant. Kentish reported that burns patients were nearly always treated in their 'own houses', often by a colliery surgeon.[99] Family members, especially female relatives, or other female household members, usually carried out general nursing duties. In the 1850s, John Wilson worked at various Durham collieries and lodged with the Dove family. When he succumbed to a 'raging fever', Wilson was nursed by his landlady, Mrs Dove, who tenderly 'watched over ... [him] night and day', and looked after him with a 'pure mother's heart'.[100] Wilson recalled Mrs Dove with fondness many years after his sickness, but home care was not always necessarily a pleasant experience. In his Sanitary Commission report on Tranent (1842), Alison stated that he had visited severely injured men whose wives were in such 'a state of intoxication' that they were actually a danger to their husbands. He claimed to have personally known of cases, for example, 'where the wife has injured the wounded husband by falling over him on the bed when she has come in' drunk.[101]

Mining families throughout Britain's coalfields, moreover, often lived in cramped conditions, unsuitable for rest or recovery. Shift work placed pressure on resources. A witness to the 1847 Commission enquiring into the state of education in Wales noted that among the houses occupied by colliers and other workers employed by the Dowlais Company, the 'sleeping-rooms are unhealthily and improperly crowded; so much so, that the beds are oftentimes occupied by relays of sleepers, who fill them two or three times successively in the 24 hours'.[102] Getting adequate or uninterrupted rest in such conditions would have been difficult. Over-crowding was reported in other areas. As a boy in the 1850s, Thomas Burt and his family lived at his cousin's house in Cramlington, Northumberland, which had only one room, but still accommodated seven to eight people while he was there.[103] There were many reports of shoddily constructed, unsanitary homes and of bedridden and injured members of mining families having to sleep under leaking roofs.[104] Other aspects of the built environment may have presented further challenges. While many miners' homes consisted of one storey, in others the upper level could only be reached by a ladder. For people with mobility impairments, such a means of ascent may have made parts of their homes inaccessible.[105] While the support of family members was crucial, then, the home environment may have presented challenges to the sick and injured in coalfield communities.

Although the home remained the most important site for the medical care of sick and disabled mineworkers in this period, institutions outside the domestic sphere did become more significant over time. The first form of specialist hospital provision for workers in the coal industry had been established in the eighteenth century by the Keelmen's Company in Newcastle, with the support of coal owners, to provide assistance for those who loaded coal onto boats on the Tyne. Established partly to address the problem of obtaining Poor Law medical services for a working population that included many migrants from Scotland who lacked legal settlement in the city, the Keelmen's Hospital provided both institutional medical care and sickness benefits paid to members in their own homes.[106] The Keelmen's Hospital opened in 1701 as an almshouse to provide a 'comfortable asylum' and source of support for all the 'aged and distressed among the keelmen' of Newcastle.[107] For others, the compulsory medical insurance schemes established by mine owners allowed access to hospital care. Mining companies and friendly societies subscribed to hospitals and used their rights as subscribers to obtain medical care for sick and injured mineworkers.[108] The records of the Glasgow Royal Infirmary reveal, for example, that collier William Preston was treated there for a fractured leg in 1856, following a recommendation for admission by Carfin Colliery. He was just one of many mineworkers admitted that year on a 'ticket' from a subscribing mining company.[109] Similar practices occurred in other British coalfields.[110]

The growing provision of specialist medical institutions, such as eye hospitals, was a feature of nineteenth-century medicine and these too accepted subscriptions from mining companies for the treatment of injured workers.[111] The Glasgow Ophthalmic Institution admitted several miners during the 1870s, performing operations such as the removal of an eye as a result of disease and injury. The ethos of the institution, in common with others, was to restore the body to usefulness, and through clinical intervention avoid patients and their families 'falling into a state of destitution and dependency'. Such institutions saw themselves as standing at the vanguard of medical efforts to prevent serious diseases or injuries becoming disabling. Without such careful interventions, the hospital's annual report noted in 1871, 'there is great danger of the patient falling into complete blindness'.[112] Nevertheless, by and large hospitals were only interested in 'curable' cases, which meant that those with disabling injuries or chronic conditions not amenable to treatment were often excluded.[113] The Glasgow Ophthalmic Institution's report for 1872 boasted that 1946 cases had been cured, forty-one were 'relieved', one patient died and 'only 57 were dismissed incurable', usually because their eyes had been 'hopelessly torn by accident or otherwise injured beyond remedy before being presented at the Institution'. The telling inclusion of the word 'only' spoke volumes about the priorities of Victorian medicine: 'incurables' represented failure or lack of hope which sat awkwardly with the faith in scientific progress that specialist institutions such as this sought to embody.[114]

Geography also affected miners' access to institutional medical care. Although most sick and injured colliers were usually treated in their own homes in the eighteenth century, those working in close proximity to an established infirmary were more likely to receive institutional care. In the 1790s, collier James Cameron was admitted to the Glasgow Royal Infirmary with an 'Ulcer' on his legs. As the institution's admission records list his parish as Glasgow, it seems Cameron lived fairly close by and was therefore able to get to the hospital relatively easily.[115] For those further afield, the distance to the nearest hospital represented a practical barrier that many could not, or would not, surmount. The nearest hospital to the mining town of Tranent, reported Alison, was reckoned to be ten miles away and inhabitants were reluctant to go there because of the 'expense and fatigue of travelling'. The hospital, moreover, apparently had a bad reputation 'among the poor classes' as it was believed patients there could not expect 'good usage' from nursing staff.[116] In the south Wales valleys at the end of the century, Edmund Stonelake painted a vivid picture of mining communities in which operations were performed on kitchen tables, with limbs removed 'just as a butcher saws a bone on his block'. There was, he wrote, 'no alternative' to this 'crude way' of practising medicine, 'as hospitals were to be found only in large towns'.[117] In spite of an expansion of hospitals and the passing of legislation in 1867 that acknowledged the duty of the state to provide hospital care for the poor, access to institutional medicine remained limited.[118] According to one estimate of hospital capacity in the 1880s, there was one bed to every 980 inhabitants in England and one to 930 in Scotland. With an estimated ratio of one bed to every 2,340 of population, Wales fared much worse.[119] Where a mineworker happened to live clearly affected his or her chances of finding a hospital bed. Moreover, during epidemics, it was not uncommon for coalfield-serving hospitals to prioritise fever patients at the expense of surgical cases.[120]

In eighteenth- and nineteenth-century England and Wales, members of the labouring poor commonly sought sanctuary and treatment in workhouses when they were unable to work because of ill health or injury. The relatively high wages earned by mineworkers sometimes disqualified them from Poor Law medical assistance, but workhouse admission registers from the coalfields show that on occasion injured mineworkers may have sought help from these institutions. On 15 May 1867, for example, forty-year-old collier Lewis Williams was admitted to the Pontypool workhouse with a broken thigh. His stay was a fairly short one, a workhouse official recording that Williams was 'discharged at [his] own request' less than three months later on 8 August. Sixteen-year-old Richard Moss was also admitted to the workhouse that year after being 'burnt in a coal pit', staying for two months, before requesting a discharge after he was 'cured of his burns'.[121] For most people, workhouse medical provision was a temporary measure and most left as soon as they were sufficiently recovered to resume life outside, often – as these cases and many more indicate – of their own volition.[122] Despite the reputation of Victorian workhouses as places of severe discipline, sick and injured paupers were frequently able to exercise some control over the duration of their institutional medical care.

By the second half of the nineteenth century, many Victorians realised that the medical infrastructure in Britain's coal-producing regions was struggling to cope with rapid urbanisation and the influx of new inhabitants to pit communities.[123] With the exception of institutions like the Keelmen's Hospital, which provided long-term treatment for the disabled and elderly workers of the Newcastle coal trade, most hospital provision was geared towards acute care. The task of rehabilitating injured workers was not a medical priority for much of our period and tended to take place on an ad hoc basis with workers' friends and family taking a prominent role. For example, when John Wilson fought to get back to normal life after his bout of incapacitating illness while living with the Dove family in north-east England in the 1850s, he did not call upon the services of medical professionals. The 'first time' he 'ventured out' after falling ill, Wilson recalled, 'I was led to the door by my good old friend, and with hands pressed to the wall (as I was not able to go without support) I managed a few yards and back, increasing strength coming with every morning's effort.'[124]

The provision of prosthetic limbs or assistive technologies to help mineworkers recover from injury, or adapt to life after amputation, was similarly ad hoc. As Thomas Burt's memory of the 'many crutches' and 'wooden legs' he noticed among workers at Murton Colliery in the 1850s indicates, mobility devices such as these were a common sight in coalfield communities, but for most of the nineteenth century injured mineworkers, their families or friends acquired or made these themselves with little, if any, help from outside agencies or organisations. Prosthetic limbs were often expensive to obtain and could be beyond the means of some amputee miners, particularly before the labour movement took up their plight in earnest in the early twentieth century.[125] In such cases, injured workers sometimes turned to the Poor Law or friendly societies for help acquiring artificial limbs. These efforts met with varying degrees of success and could depend on the circumstances of dismemberment or whether or not the impaired person was likely to be able to return to work and reimburse the authorities for her or his 'loan'.[126]

Although injured industrial workers commonly used assistive devices throughout the nineteenth century, the provision of such technology can hardly be regarded as systematic before the twentieth century. While much rehabilitation work took place within families and communities, a step towards institutionalised rehabilitative medicine for workers in one of Britain's most risk-prone coalfields was taken at the end our period with the establishment of the Rest convalescent home on the south Wales coast. Founded in 1862 by Dr James Lewis, medical officer of the Bridgend and Cowbridge Poor Law Union, the Rest was originally intended for the 'Invalid Poor' of Glamorganshire and surrounding areas. Given the importance of mining to the local economy, however, the institution soon gained a reputation as a place primarily for ill or injured miners and they made up a majority of its residents. Offering patients a therapeutic regimen that made the most of its seaside location and the supposedly recuperative benefits of the coastal climate and sea bathing, the home initially operated out of three cottages but eventually moved to new purpose-built premises in Porthcawl in 1878. During its first year of operation there the Rest took in thirty-three patients. By the early twentieth century it was regularly housing more than 1000 a year.[127]

Supporters of the Rest and its enlargement saw it as a 'convalescent ward' (or a 'handmaid to the hospitals', as one writer put it) designed to support the work of 'local infirmaries'. The goal of the institution was to provide those 'not sufficiently recovered to enter upon their daily labours' with everything they needed to regain their health properly to enable a sustained return to work. Rush them back to ordinary life too early, it was argued, and recovering ill or injured workers might relapse into incapacity or, worse still, slip into permanent 'debility'. The Rest, then, was not meant for permanently disabled workers, but rather those who were in a liminal state somewhere between health and illness or disability and 'able-bodiedness'. Applicants for residence were only accepted if they had a doctor's certificate testifying that there was a good chance 'treatment' in the home would significantly aid their recovery. Residents, moreover, were usually only allowed to stay for a few weeks at most. Those requiring longer-term care were generally not welcome.[128] In the final analysis, the Rest was more about preventing long-term incapacity than managing it. Like all the medical interventions in miners' lives described in this chapter, its fundamental goal was to help get injured workers back to work.[129]

'Medicalisation', conflict and authority

While medical care remained patchy throughout the period in question, there can be no doubt that the expansion of medical services in response to the perceived health risks of coalmining was a significant feature of the industry's expansion in the century after 1780. It is likely that some sick and injured miners benefitted significantly from access to medical professionals and that, ultimately, greater scientific interest in the 'habits and diseases' of miners, especially lung disease, would lead in the long run to improved therapies and interventions.[130] Nevertheless, as we have seen, there was much scope for antagonism around sensitive issues such as dissection. While miners' continued resort to unorthodox healers, or indeed local chemists and druggists, might have proceeded in some cases more from necessity than free choice, doctors' criticism of mineworkers' 'superstition' and their use of time-worn remedies suggests that sick and disabled mineworkers were willing to exercise some agency when it came to medical care.[131] What, then, was the relationship between mineworkers and doctors?

Relations between doctors and patients in the eighteenth and nineteenth centuries have been a topic of considerable debate. Medical historians have seen the late eighteenth and early nineteenth centuries as marking a fundamental shift in medical knowledge and power which amounted to the growth of professional authority over medical matters and the gradual 'de-skilling' of lay people as interpreters of their own health and illness. Factors such as the rise of pathological anatomy, the increasing use of specialist diagnostic technology (epitomised by Laennec's invention of the stethoscope) and the shift from viewing illness in holistic terms as a disruption to the individual patient's constitution towards diagnosing it in terms of a series of universal symptoms, are all cited as evidence of the diminishing power of patients to challenge medical authority.[132] And although the public health movement attracted relatively few doctors in its first stages, there can be no doubt that public investigations into the health of the industrial population increased the profile of medical men in public life as 'experts'.[133] The increasing importance of medical professionals and paradigms in the identification and treatment of physiological 'disorders' and in pressing for reform is often seen as evidence of the 'medicalisation of society'.[134] A similar analysis informs ideas about the 'medical model' of disability, which in disability studies is seen as gathering pace from the late eighteenth century. During this period, Paul Longmore has argued, there was a shift from a model of disability as an 'immutable condition caused by supernatural agency' to one which 'redefined it as a biological insufficiency amenable to professional treatment'. Doctors thus came to wield increasing power over disabled people because of their growing role in the classification, examination and treatment of disability.[135]

Nevertheless, historians have challenged the view of nineteenth-century patients as 'servile acceptors of medical orthodoxy'.[136] Studies of medical practice have shown how for much of the nineteenth century, professional authority was far from hegemonic.[137] From a disability perspective, recent work has also challenged a rapid and wholesale shift in the 'medicalisation' of impairment during the late eighteenth and nineteenth centuries. Hospitals' widespread rejection of 'incurable' cases during the period, for example, indicates that medical practitioners were not especially interested in permanent disability from a professional standpoint.[138] Evidence from Britain's coalfield communities supports this more complicated picture of conflict and negotiation in medical care.

James C. Riley has described a mutual distrust between doctors and working-class patients in nineteenth-century Britain.[139] Supporting the notion, explored in the previous chapter, that miners contributed to accidents through their 'recklessness' was an enduring belief that miners were responsible for their own ill health through their inappropriate behaviours and lifestyles. A moralistic model of sickness and disability co-existed with more objective diagnoses of symptoms and was reinforced by the rhetoric of 'habits and diseases' central to Victorian social and medical investigation.[140] In 1844 Dr James Black expressed a common opinion when he blamed miners' 'spasmodic complaints' on their 'intemperance' rather than 'from any special causes attending their employment'.[141] Such sentiments show how the corporeal objectives of medicine were often wedded to moral goals, just as they had been in earlier periods.[142] These attitudes could affect the treatment that injured miners received from doctors in quite profound ways. Nineteenth-century club doctors occupied a powerful position as 'gatekeepers' to medical services and were sometimes known to refuse treatment to miners if they suspected an injury had been caused by excessive drinking. In August 1863 William Edwards of Oakengates in Shropshire alerted readers of the *Miner and Workmen's Advocate* to the case of a collier who had received a potentially disabling blow to his ankle at work. In spite of obtaining a note from the surgeon who attended the accident to prove that his 'ankle had been crushed', the club doctor refused to attend him because he said that it 'came through drinking', hence the man was left to go to the infirmary under his own volition. 'So this poor man was swindled out of his due', wrote Edwards, 'after having paid 35 years, to be cast on the world without his pay or medical assistance to which he had a right.'[143]

Medical practitioners expected patients to defer to their expertise and follow their advice without complaint.[144] Working-class patients, however, often had other ideas and behaved in ways that fell far short of their doctors' hopes, or even challenged practitioners and their methods outright. Medical men were evaluated by their patients according to their skill and attentiveness to their duties.[145] While criticism of medical practitioners and their methods may have been most vocal in mid-Victorian friendly societies, where they often found their treatments and diagnoses challenged at meetings in which members 'adopted an independent and sometimes insolent attitude towards their superiors', the contesting of medical authority was evident in other contexts throughout our period.[146] As Kentish made clear in his *Essay on Burns*, encounters between medical men and miners usually involved a process of negotiation between the two parties. Kentish believed burnt miners' wounds should only be dressed once a day. More than this was unnecessary and potentially disturbed the patient from much needed rest. Kentish accepted, though, that implementing his recommendation was not always practically possible in mining communities because of the 'prejudices of the patient and his friends'. In such circumstances, Kentish advised, it was probably better to give in to the wishes of miners and their families and dress the wound a second time.[147] Compromise was at the heart of doctor–patient relations in the coalfields, especially before the mid-nineteenth century when the medical profession was still struggling to establish its authority. Although Kentish thought he knew best, he realised his patients often had a different opinion and that it was sometimes necessary to accede to their 'prejudices' if he was to remain their surgeon.

On issues of greater severity, such as amputation, the resistance of injured miners to the advice of their doctors could be even more determined, and understandably so given the risks of surgery at this time. During the late eighteenth century, amputation was a surgical procedure that demonstrated the power of hospital-trained surgeons over the bodies of their patients and one that aroused resistance. In 1794 the case of a collier who recovered from a compound fracture after he had 'refused to be removed to the county infirmary or submit to an amputation', provided 'striking proof of the necessity there is for great deliberation in cases where amputation may be thought necessary'.[148] The case was one of several well-publicised examples in which patients had successfully recovered after resisting surgical advice to amputate, illustrating the hastiness by which such operations were sometimes advised.[149] The issue continued to prove controversial during the nineteenth century, especially in mining where fractures were common. Giving evidence to the 1842 Children's Employment Commission, William Morison recalled the case of a sixteen-year-old boy at Newbottle Colliery who had died after suffering a compound fracture of his thigh and leg because his parents had 'resisted amputation'. A thirteen-year-old working at Sacristan Colliery had likewise died following amputation because the operation had been 'resisted until too late'.[150]

For Morison, such resistance was further evidence of the ignorance of miners in the face of medical knowledge and chimed with his comments on miners' 'superstition', but many feared a dangerous operation that, if they survived, would leave them with a permanent impairment. Improvements in techniques in the second half of the nineteenth century may have increased the pressure on patients to submit to surgical interventions, but did not eradicate the dangers of surgery.[151] In September 1862 *The Merthyr Telegraph* reported the 'Death under Chloroform' of a young 'cripple' named Henry Davies, who had injured his knee at Middle Duffryn Colliery a few years previously. Following consultations with 'several medical gentlemen' who were of 'unanimous opinion that a portion of the bone would have to be removed before the boy would recover the use of his leg', Davies was anaesthetised and 'prepared' for surgery. Before the procedure could be performed, however, the unfortunate teenager 'expired under their hands'.[152] Such negative publicity ensured resistance to amputation among injured mineworkers continued to be noted into the twentieth century.[153]

On rare occasions, injured miners who believed they had been badly treated by doctors were prepared to seek legal redress. In August 1835, the *Cambrian* newspaper reported the case of twenty-three-year-old Michael Regan, a miner whose hip had been so badly damaged in an accident that he had become 'a miserable cripple for life', unable to support himself 'by his own exertions'. After the accident, Regan was attended by surgeon 'Mr Russell, his assistants, and [an] apprentice'. Despite Regan's repeated complaints, his medical attendees dismissed the idea that he had seriously hurt his hip and left the injury untreated. In consequence of their incorrect assessment, Regan and his representatives argued, his long-term condition was worse than it would have been had he received proper treatment. In effect, Regan was blaming his disability on the incompetence of Russell and his assistants. That he did so suggests that notions about where responsibility for disability resided could vary quite considerably between patients and medical practitioners. Just as doctors were prepared to cite the actions of working people as a major cause of disability, workers like Regan instead blamed their impairments on doctors' incompetence. Regan won his case and was awarded damages of £25.[154] In a similar case in Lancashire in 1861, a miner left 'crippled' after his colliery surgeon had mistakenly diagnosed his fractured knee as merely dislocated was helped by his friends to bring a successful action for damages, leading to an award of £45.[155]

These cases, though uncommon, raised questions not just about the proper care an individual injured miner should have received, but about the broader expectations of those who subscribed to pit clubs and other workplace schemes for thorough and competent medical care. The defendant in the Regan case was described as a surgeon 'of considerable standing at Merthyr, and the opulent incumbent of a medical benefice consisting of several Iron Works, having a population of 12,000 or 13,000 persons' who each subscribed to a company medical scheme. In return for their subscriptions they felt entitled to a level of surgical and medical care whereby 'the poorest of them' should receive the 'same kind of care, the same patient assiduity' as that received by 'wealthier persons, who would pay him for each visit'.[156] Payment into a company scheme, whether compulsory or not, made some miners feel entitled to speak out against practitioner neglect. Writing under the pseudonym 'Cumro Bach', a correspondent to the *Miner and Workmen's Advocate* complained that one of his fellow workers at Nantyglo Colliery in Monmouthshire had died following an accident, where he was struck by machinery, after the colliery doctor had neglected to visit him three times after being called. However, the story was fiercely refuted by J. H. Wood, assistant surgeon to the Nantyglo Ironworks Company, who instead blamed the man for waiting too long before calling the doctor and the paper was forced to condemn 'Cumro Bach' for his misinformation.[157]

Although fiercely contested, Cumro Bach's criticisms illustrate broader tensions between miners and company-appointed medical attendants. Miners in some collieries were balloted on the appointment of a doctor, which gave them some control over how their subscriptions to sick clubs were spent, but in many cases the appointment of doctors lay in the hands of mine owners.[158] Ultimately dependent on the approval of owners for their positions, many colliery doctors seem to have felt a pressure to ensure injured miners returned to work as quickly as possible, so as to maintain a good relationship with owners. That at least was the impression many colliers had. Doctors were frequently suspected of a clash of interests between their own career ambitions and their duty of care to their patients that potentially put the welfare of sick or injured miners at risk.[159] By the later part of the period, the matter was becoming increasingly politicised thanks to the National Association of Coal Miners. The union criticised pit clubs for creating a 'large amount of capital for the use of the employer', for which balance sheets were rarely made public. What was more, the men were obliged to have the 'coalowners' nominee for a medical attendant', such that 'the colliers are doctored by contract at their own expense for the benefit of their employers'.[160] At South Dunraven Colliery near Treherbert in 1886, miners opposed the withholding of their poundage money 'against their will' to employ one Dr Warburton, a man 'whose services the great majority of them do not want to retain'. The issue proved a test of the powers of employers, with one prominent barrister arguing that 'every penny kept back from the men's wages towards the payment of doctors without the men's special written consent is illegal'.[161]

Conclusion

The histories of British coalmining and medicine are closely interwoven. During the century from 1780 to 1880, coalmining and mineworkers helped shape medicine and the emerging relationship between medical practitioners and working-class patients. The 'habits and diseases' of miners became a topic of increasing public scrutiny as the industry expanded, thanks to innovations in public health and evolving research into conditions such as lung diseases. The effects of mine work on different aspects of miners' bodily health were documented by medical practitioners and by those who contributed to the great parliamentary investigations into mining, such as the 1842 Children's Employment Commission, which revealed a plethora of illnesses and deformities related to underground work. While evidence of mineworkers' diseases and disabilities helped propel the movement for industrial reform, the image of the miner's body presented in the statements of witnesses to these parliamentary inquiries was contradictory, being simultaneously admired as the epitome of muscular able-bodiedness while also distinctive for its stunted growth, 'crippled gait' and sallow complexion.[162] Likewise, colliers were distinguished for their physical prowess as a 'picked body of men', whose strength derived from a process of natural selection in coalfield communities in which the fittest were chosen for the hardest of tasks, yet simultaneously stigmatised as a 'race of men' recognisable by their presumed low intellectual capabilities and brutish insensibility to pain.

'Disease' and 'disability' overlapped in medical perceptions of the health of miners. Physical 'deformity' attributed to the posture in which colliers worked was believed not only to have produced lasting impairment, but also to have contributed to the incidence of diseases, from breathing difficulties to kidney problems. Conversely, medical case studies of those suffering from 'black lung' highlighted by medical investigators in the 1830s, reveal the progressively disabling consequences of respiratory illness, leading to a diminution of physical capabilities and increasing reliance on others. While some hospital care was provided for the long-term sick and disabled, in workhouses or through self-help schemes such as the innovative Keelmen's Hospital in Newcastle, much medical provision was geared towards acute rather than chronic conditions.

There was little place for the 'incurable' in the celebratory narratives of medical progress and surgical authority in Victorian medicine. Yet the foundation of the Rest in Porthcawl to provide an alternative to home care for those in need of recuperation acknowledged at least that the journey from sickness to wellness had a number of distinctive stages that demanded different types of intervention. Recuperative medicine, like other medical fields, aimed to stop injuries and impairments becoming disabling, and the principles behind The Rest informed subsequent efforts to restore workers' health that coalesced into the rehabilitation movement of the twentieth century.[163]

One important consequence of the expansion of coalmining in Britain after 1780 was the increasingly prominent role of the medical 'expert' in the daily lives of coal workers. Making medical provision for workers via the employment of surgeons to tend men in the wake of accidents, the provision of workplace 'sick clubs' and paying subscriptions to allow injured miners access to hospital care was praised as evidence of employer 'liberality' and sympathetic paternalism. However, it also served a political purpose by encouraging workers to remain loyal to their employers (having invested their wages in a compulsory sick club). Many miners and their families benefited from the expansion of medical attention. But this chapter has also highlighted areas where the authority of medical practitioners could be called into question. These included choosing unorthodox healers over learned practitioners, resisting dissection or refusing dangerous and disabling procedures such as amputation and even taking legal action where medical negligence had caused permanent impairment rather than cure. Subscribing to a colliery sick club made some members feel entitled to draw on services as they saw fit and to speak out against practitioners they felt were not performing their duties properly. Increasingly, this led to tensions between employers and workers over the appointment of colliery doctors.

Ultimately, concerns over the abuses of colliery sick clubs may have given powerful impetus to some miners, like other industrial workers, to seek alternative sources of medical aid, such as those provided through friendly societies.[164] Mineworkers wanted greater control over the treatment they received as well as more say in who treated them. The famous mutualism of coalfield communities was driven, in large part, by this goal and is an indication of just how politicised workers' healthcare was in industrialising Britain. The fusion of medical and financial aid in friendly society schemes also reminds us that the provision of medical treatment was bound up with broader questions of welfare. And it is to the non-medical assistance provided for sick and injured miners and their families that we now turn.

Notes

1. S. Scott Alison, 'On the Sanitary Condition and General Economy of the Town of Tranent, and the Neighbouring District in Haddingtonshire', PP 1842 (008), *Sanitary Inquiry: Scotland: Reports on the Sanitary Condition of the Labouring Population of Scotland, in Consequence of an Inquiry Directed to be Made by the Poor Law Commissioners*, 103.
2. For example, *Hansard, HC Deb*, 22 June 1842, vol. 64, cols 423–8.
3. J. Arkless, 'What are Miners?', *The Miner and Workmen's Advocate*, 16, 20 June 1863, 6.
4. PP 1842 (381), *Appendix to the First Report of the Commissioners. Mines. Part 1. Reports and Evidence from Sub-Commissioners*, 621.
5. 'The Collieries – No. 1', *Monthly Supplement of the Penny Magazine of the Society for the Diffusion of Useful Knowledge*, 28 February –1 March 1835, 123.
6. J. T. Arlidge, *The Hygiene, Diseases and Mortality of Occupations* (London: Percival and Co., 1892), 261.
7. Anthony S. Wohl, *Endangered Lives: Public Health in Victorian Britain* (London: J. M. Dent, 1983), 278.
8. Arlidge, *Hygiene*, 271.
9. PP 1842 (381), 609.
10. Edwin Chadwick, *Report on the Sanitary Condition of the Labouring Population of Great Britain 1842*, ed. M. W. Flinn (Edinburgh: Edinburgh University Press, 1965), 316; but cf. William I. Cox, 'Diseases of Special Occupations. No. III. Diseases of Colliers in South Lancashire', *British Medical Journal*, 11 July 1857, 579.
11. PP 1842 (382), 622. See also PP 1842 (381), 143, 637.
12. Arlidge, *Hygiene*, 270, 64.
13. The term appears in the 1842 Children's Employment Commission Report: PP 1842 (380), *Children's Employment Commission. First Report of the Commissioners. Mines*, 182–3.
14. Charles Turner Thackrah, *The Effects of the Principal Arts, Trades and Professions and of Civic States and Habits of Living, on Health and Longevity* (London: Longman and others, 1831), 112.
15. 'The Collieries', 123.
16. Douglas A. Lorimer, *Colour, Class, and the Victorians: English attitudes to the Negro in the Mid-Nineteenth Century* (Leicester: Leicester University Press, 1978); Alexandra Jones, 'Disability in Coalfields Literature c. 1880–1948: A Comparative Study' (unpublished PhD thesis, Swansea University, 2016), 83–90.
17. PP 1842 (381), 525; J. R. Leifchild, *Our Coal and our Coal Pits* (1853; London: Frank Cass, 1968), 197.
18. S. Scott Alison, 'On the Diseases, Conditions, and Habits of the Collier Population of East Lothian', PP 1842 (381), 411.
19. G. Mallett, 'Remarks on Accidents Occurring to Colliers', *Association Medical Journal*, 12 October 1855, 930.

20 Ibid., 929, 930. Our italics. Joanna Bourke, *What it Means to be Human: Reflections from 1791 to the Present* (London: Virago, 2011), 78–81, 105; Joanna Bourke, *The Story of Pain: From Prayer to Painkillers* (Oxford: Oxford University Press, 2014), 203. See also, Lucy Binding, *The Representation of Bodily Pain in Nineteenth-Century English Culture* (Oxford: Oxford University Press, 2000), 177–239; Martin S. Pernick, *A Calculus of Suffering: Pain, Professionalism and Anaesthesia in Nineteenth-Century America* (New York: Columbia University Press, 1985).

21 George Rosen, *The History of Miners' Diseases: A Medical and Social Interpretation* (New York: Schuman's, 1943), 29.

22 Rosen, *History*, 93–4; P. W. J. Bartrip, *The Home Office and the Dangerous Trades: Regulating Occupational Disease in Victorian and Edwardian Britain* (Amsterdam: Rodopi, 2002), 12–13; Giuliano Franco and Francesca Franco, 'Bernardino Ramazzini: The Father of Occupational Medicine', *American Journal of Public Health*, 91 (2001), 1382.

23 Peter Kirby, *Child Workers and Industrial Health in Britain 1780–1850* (Woodbridge: The Boydell Press, 2013), 63.

24 Thackrah, *Effects*, 106.

25 PP 1842 (381), 665, 673.

26 William I. Cox, 'Diseases of Colliers in South Lancashire', *British Medical Journal*, 11 July 1857, 579.

27 Beth Linker, 'On the Borderland of Medical and Disability History: A Survey of the Fields', *Bulletin of the History of Medicine*, 87:4 (2013), 499–535.

28 Rosen, *History*, 137. For example, see William Buchan, *Domestic Medicine: Or, a Treatise on the Prevention and Cure of Diseases by Regimen and Simple Medicines*, Eleventh Edition (London: A. Strahan et al., 1790), 38.

29 Edward Kentish, *An Essay on Burns, Principally upon Those Which Happen to Workmen in Mines* (London: 1797), 3.

30 Ibid., 4.

31 Ibid., 3.

32 Robert Bald, *A General View of the Coal Trade of Scotland* (Edinburgh: Oliphant, Waugh and Innes, 1812), 130, 131, 138.

33 Rosen, *History*, 191, 260, 298. For the recognition that geological conditions made lung disease particularly prevalent in the mines of Fifeshire and Midlothian see: George Steele, 'Of the Expectoration of Black Matter from the Lungs', *London Medical Gazette, or Journal of Practical Medicine*, 20 (1837), http://www.scottishmining.co.uk/441.html, accessed 21 July 2014; W. Sanders, 'Observations on Some of the Objects of Interest Contained in the Museum of the College of Surgeons of Edinburgh', *British Medical Journal*, 14 August 1858, 676.

34 Rosen, *History*, 244–401, presents an extensive history of medical progress made in the identification of lung diseases in miners; see also Andrew Meiklejohn, 'History of Lung Diseases of Coal Miners in Great Britain: Part I, 1800–1875', *British Journal of Industrial Medicine*, 8 (1951), 127–37; Arthur McIvor and

Ronald Johnston, *Miners' Lung: A History of Lung Disease in British Coal Mining* (Aldershot: Ashgate, 2007), 64–8.
35 William Thomson, 'On Black Expectoration, and the Deposition of Black Matter in the Lungs, Particularly as Occurring in Coal Miners etc', *Medical Chirurgical Transactions*, 20 (1837), 230–300.
36 Ibid., 249–50.
37 Ibid., 258.
38 Ibid., 241, 273. We are grateful to Evelien Bracke for advice on translation.
39 Ruth Richardson, 'Popular Beliefs about the Dead Body', in Carole Reeves (ed.), *A Cultural History of the Human Body in the Age of Enlightenment* (Oxford: Berg, 2010), 106; Ruth Richardson, *Death, Dissection and the Destitute* (Chicago and London: Chicago University Press, 2000), 131–58, 193–7, 259.
40 Christopher Lawrence, *Medicine in the Making of Modern Britain 1700–1920* (London and New York: Routledge, 1994), 23.
41 Robert Gray, 'Medical Men, Industrial Labour and the State in Britain, 1830–50', *Social History*, 16:1 (1991), 28.
42 John C. Cobden, *The White Slaves of England. Compiled from Official Documents* (Auburn and Buffalo, NY: Miller, Orton and Mulligan, 1854), 67.
43 PP 1842 (381), 153.
44 Ibid., 655.
45 Ibid., 573.
46 Ibid., 643.
47 Ibid., 626.
48 Cobden, *White Slaves of England*, 39, 56, 60, 61, 62.
49 Cited in ibid., 80.
50 PP 1842 (381), 387.
51 Ibid., 366.
52 PP 1842 (382), 625.
53 Cobden, *White Slaves of England*, 85.
54 PP 1842 (381), 642.
55 J. B. Thomson, 'The Melanosis of Miners; or Spurious Melanosis', *Edinburgh Medical Journal*, 4 (1858), 226–7, 228.
56 Rosen, *History*, 419.
57 James C. Riley, *Sick, Not Dead: the Health of British Workingmen during the Mortality Decline* (Baltimore: Johns Hopkins University Press, 1997), ch. 2; David G. Green, *Working-Class Patients and the Medical Establishment: Self-Help in Britain from the Mid-Nineteenth Century to 1948* (Aldershot: Gower, 1985). The welfare role of friendly societies is discussed more fully in Chapter 3.
58 Herbert Mackworth, 'Mines: Accidents in Them, and Sanitary Condition of Them' in *Lectures Delivered at the Bristol Mining School, 1857* (Bristol: Bristol Mining School, 1859), 194.
59 Anne Digby, *The Evolution of British General Practice, 1850–1948* (Oxford: Oxford University Press, 1999), 271–5; Nigel Nauton Davies, 'Two and a Half

Centuries of Medical Practice: A Welsh Medical Dynasty', in John Cule (ed.), *Wales and Medicine: An Historical Survey from Papers Given at the Ninth British Congress on the History of Medicine* (Llandysul: Gomer, 1975), 216–21.
60 Ibid., 217.
61 PP 1842 (381), 150.
62 Ibid., 554.
63 Digby, *Evolution of British General Practice*, 272.
64 PP 1842 (381), 664.
65 Ibid., 645.
66 Ray Earwicker, 'Miners' Medical Services before the First World War: the South Wales Coalfield', *Llafur*, 3:2 (1981), 40.
67 PP 1842 (381), 645.
68 Ibid., 568.
69 Earwicker, 'Miners' Medical Services', 40.
70 PP 1842 (382), 710.
71 PP 1842 (382), 548.
72 Ibid., 640; Earwicker, 'Miners' Medical Services', 42.
73 For example, Glasgow University Archives, UGD1/37/1, Govan Colliery Paybook November 1855; UGD/1/37/12 Govan Colliery Pay Book, June 1862–June 1863.
74 'The Collier at Home', *Household Words*, 15: 366 (28 March 1857), 290.
75 Anthony Errington, *Coals on Rails, or the Reason of my Wrighting: the Autobiography of Anthony Errington, a Tyneside Colliery Waggon and Waggonway Wright from his Birth in 1778 to around 1825*, ed. P. E. H. Hair (Liverpool: Liverpool University Press, 1988), 63. For other examples, see James Everett, *The Wall's End Miner; or, a Brief Memoir of the Life of William Crister Including an Account of the Catastrophe of June 18th 1835* (London: Hamilton, Adams and Col, 1835), 135–16, 161; George Parkinson, *True Stories of Durham Pit-Life* (London: C. H. Kelly, 1912), 51.
76 Mackworth, 'Mines', 194.
77 Ibid.
78 'Palanquins for the Wounded', *Colliery Guardian and Journal of the Coal and Iron Trades*, 12 June 1858, 372.
79 Edmund Stonelake, *The Autobiography of Edmund Stonelake*, ed. A. Mór-O'Brien (Bridgend: D. Brown and Sons, 1981), 95; Jamie L. Bronstein, *Caught in the Machinery: Workplace Accidents and Injured Workers in Nineteenth-Century Britain* (Stanford, CA: Stanford University Press, 2008), 60–3, 90.
80 'Palanquins for the Wounded', 372.
81 'South Staffordshire and East Worcestershire Institute of Mining Engineering. Important Communication upon the Treatment and Removal of Accident Cases', *Colliery Guardian and Journal of the Coal and Iron Trades*, 11 June 1875, 855. See also 'Portable Ambulance for Miners', *Colliery Guardian and Journal of the Coal and Iron Trades*, 8 March 1878, 389; Roger Cooter, 'The Moment of the

Accident: Culture, Militarism and Modernity in Late-Victorian Britain' in Bill Luckin and Roger Cooter (eds), *Accidents in History: Injuries, Fatalities and Social Relations* (Amsterdam and Atlanta GA: Rodopi, 1997), 107–57; P. W. J. Bartrip and S. B. Burman, *The Wounded Soldiers of Industry: Industrial Compensation Policy 1833–1897* (Oxford: Clarendon Press, 1983), 5; Bronstein, *Caught in the Machinery*, 74.

82 Roy Porter, 'Accidents in the Eighteenth Century' in Cooter and Luckin (eds), *Accidents in History*, 96.
83 Digby, *Evolution of British General Practice*, 274.
84 *Sunday Chronicle*, 23 September 1787.
85 'The Hartley Colliery Calamity', *Aberdare Times*, 9 February 1862.
86 Mackworth, 'Mines', 194.
87 'South Staffordshire and East Worcestershire Institute of Mining Engineering', 855.
88 Errington, *Coals on Rails*, 35. Shared belief in the efficacy of humoral treatments was common in both 'orthodox' and irregular healers treating mining cases, such as burns, in the eighteenth century: see Kentish, *Essay on Burns*, 82.
89 PP 1842 (382), 594.
90 PP 1842 (381), 728.
91 PP 1842 (381), 667.
92 Peter Bartrip, *Themselves Writ Large: the British Medical Association 1832–1866* (London: British Medical Journal, 1996), 97.
93 Stonelake, *Autobiography*, 177.
94 *The Miner and Workmen's Advocate*, no. 15, 13 June 1863.
95 F. B. Smith, *The People's Health, 1830–1910* (London: Croom Helm, 1979), 266.
96 Kentish, *Essay on Burns*, 75, 78–9, 86, 93–99, 148.
97 Bourke, *Story of Pain*, ch. 9.
98 Everett, *Wall's End Miner*, 136–41.
99 Kentish, *Essay on Burns*, 72.
100 John Wilson, *Autobiography of John Wilson, J.P., M.P. ... Reprinted from 'The Durham Chronicle'* (Durham: Durham Chronicle, 1909), 29.
101 PP 1842 (008), 99–100.
102 *Reports of the Commissioners of Inquiry into the State of Education in Wales, Appointed by the Committee of Council on Education, in Three Parts. Part 1. Carmarthen, Glamorgan and Pembroke* (London: William Clowes and Sons for HMSO, 1847), 484.
103 Thomas Burt and Aaron Watson, *Thomas Burt, M.P., D.C.L, Pitman and Privy Councilor: An Autobiography, with Supplementary Chapters by Aaron Watson* (London: Unwin, 1924), 91.
104 John Benson, *British Coalminers in the Nineteenth Century: A Social History* (Dublin: Gill and Macmillan, 1980), 93–103; Robert Duncan, *The Mineworkers* (Edinburgh: Birlinn, 2005), 129–30; PP 1842 (008), 129.

105 Benson, *British Coalminers*, 96.
106 Tyne and Wear Archives, CH.KH/1 Keelmen's Hospital, Newcastle upon Tyne, minutes 7 January 1739/40 to 17 December 1842; Joseph M. Fewster, *The Keelmen of Tyneside: Labour Organisation and Conflict in the North-East Coal Industry* (Woodbridge: Boydell Press, 2011), 48.
107 *Articles of the Keelmen's Hospital Society; with Rules and Regulations for the Hospital* (Newcastle upon Tyne: John Marshall, 1829), 22.
108 Anne Borsay, *Disability and Social Policy in Britain since 1750: A History of Exclusion* (Basingstoke: Palgrave Macmillan, 2005), 46–9.
109 NHS Greater Glasgow and Clyde Archives, HH67/56/20 Glasgow Royal Infirmary Admission and Dismission Register January 1856 to November 1857; Smith, *The People's Health*. See also *Statistical Compendium* table 4.2, Glasgow Ophthalmic Institution, Register of Indoor Patients, 1876–78. http://doi.org/10.5281/zenodo.183686, accessed 24 March 2017.
110 For example, Gwent Archives, D3293/A/1, Newport and Monmouthshire Hospital Annual Reports, 1854, 7.
111 Luke Davidson, '"Identities Ascertained:" British Ophthalmology in the First Half of the Nineteenth Century', *Social History of Medicine* 9 (1996), 313–33; NHS Greater Glasgow Health Board Archives, HB47/2/1, Glasgow Ophthalmic Institution, Annual Reports, Fourth Annual Report … 3 March 1873.
112 Ibid., Report by the Directors of the Glasgow Ophthalmic Institution … 6 March 1871.
113 Smith, *The People's Health*, 252–9; Jason Szabo, *Incurable and Intolerable: Chronic Disease and Slow Death in Nineteenth-Century France* (New Brunswick NJ: Rutgers University Press, 2009).
114 NHS Greater Glasgow Health Board Archives, HB47/2/1, Glasgow Ophthalmic Institution, Annual Reports, Third Annual Report by the Directors of the Glasgow Ophthalmic Institution, 4 March 1872.
115 Greater Glasgow and Clyde NHS Archives, HH67/56/1A, Glasgow Royal Infirmary, Admission and Dismission Register, 1794–1800.
116 PP 1842 (008), 90.
117 Stonelake, *Autobiography*, 95.
118 M. W. Flinn, 'Medical Services Under the New Poor Law' in Derek Fraser (ed.), *The New Poor Law in the Nineteenth Century* (Basingstoke: Macmillan, 1976), 65.
119 Smith, *The People's Health*, 251.
120 Marguerite W. Dupree, 'Family Care and Hospital Care: the 'Sick Poor', in Nineteenth-Century Glasgow', *Social History of Medicine* 6 (1993), 195–211.
121 Gwent Archives, CSW/BGP/I/224, Pontypool Union, Workhouse Admission and Discharge Books, 1865–67; CSW/BGP/I/225, Pontypool Union, Workhouse Admission and Discharge books, 1867–69. For examples from the 1870s, see CSW/BGP/I/226, Admission and Discharge Books, 1869–72 and CSW/BGP/I/227, Admission and Discharge Books, 1872–74.
122 Marjorie Levine-Clark, 'Engendering Relief: Women, Ablebodiedness, and the

New Poor Law in Early Victorian England', *Journal of Women's History*, 11:4 (2000), 121.
123 For example, Gwent Archives, D3293/A/1, *Newport and Monmouthshire Hospital Annual Report*, 1854, 15.
124 Wilson, *Autobiography*, 29.
125 Ben Curtis and Steven Thompson, '"A Plentiful Crop of Cripples Made By All this Progress:" Disability, Artificial Limbs and Working-Class Mutualism in the South Wales Coalfield, 1890–1948', *Social History of Medicine*, 27 (2014), 708–27.
126 See, for example, Tyne and Wear Archives, PU.SS/1/1/19 South Shields Poor Law Union Board of Guardians Minute Book, 1863–65, correspondence concerning William Butler's artificial leg, 34, 44, 50.
127 *Statistical Compendium* table 5.12, The Rest Convalescent Home, Porthcawl, Admissions, 1878–1938, http://doi.org/10.5281/zenodo.183686, accessed 24 March 2017.
128 'The Rest, Porthcawl', *The Cardiff Times*, 25 November 1871; *South Wales Daily News*, 19 November 1877; *South Wales Daily News*, 19 November 1877; 'The Porthcawl "Rest"', *Western Mail*, 24 March 1875; *The Cardiff Times*, 8 April 1876; 'Porthcawl Rest', *Western Mail*, 15 May 1879; Hilary M. Thomas, *The Rest (founded 1862): A Brief History* (Port Talbot: D. W. Jones, 1988).
129 Anne Borsay, 'Returning Patients to the Community: Disability, Medicine and Economic Rationality Before the Industrial Revolution', *Disability and Society*, 13:5 (1998), 645–63.
130 McIvor and Johnston, *Miners' Lung*, 65 and *passim*.
131 On the importance of the chemist's shop in the medical landscape of coalfield communities see Stonelake, *Autobiography*, 36–7.
132 Lawrence, *Medicine in the Making of Modern Britain*, 22–3; Mary E. Fissell, 'The Disappearance of the Patient's Narrative and the Invention of Hospital Medicine' in Roger French and Andrew Wear (eds), *British Medicine in the Age of Reform* (London and New York: Routledge, 1991), 92–109; Mary E. Fissell, *Patients, Power and the Poor in Eighteenth-Century Bristol* (Cambridge: Cambridge University Press, 1991); Michel Foucault, *The Birth of the Clinic: An Archaeology of Medical Perception* (New York: Vintage Books, 1973); Stanley Joel Reiser, *Medicine and the Reign of Technology* (Cambridge: Cambridge University Press, 1998); N. D. Jewson, 'The Disappearance of the Sick-Man from Medical Cosmology, 1770–1870', *Sociology*, 10 (1976), 225–44.
133 Lawrence, *Medicine in the Making of Modern Britain*, 40–44; Michael E. Rose, 'The Doctor in the Industrial Revolution', *British Journal of Industrial Medicine*, 28 (1971), 22–6.
134 Peter Conrad, *The Medicalization of Society: On the Transformation of Human Conditions into Treatable Disorders* (Baltimore: Johns Hopkins University Press, 2007).
135 Paul K. Longmore, *Why I Burned My Book and Other Essays on Disability*

(Philadelphia: Temple University Press, 2003), 41; Lennard J. Davis, 'Dr Johnson, Amelia and the Discourse of Disability in the Eighteenth Century' in Helen Deutsch and Felicity A. Nussbaum (eds), *'Defects': Engendering the Modern Body* (Ann Arbor: University of Michigan Press, 2000), 56; Dwight Christopher Gabbard, 'Disability Studies and the British Long Eighteenth Century', *Literature Compass*, 8 (2011), 83–4.

136 Michael Neve, 'Orthodoxy and Fringe: Medicine in Late Georgian Bristol' in W. F. Bynum and Roy Porter (eds), *Medical Fringe and Medical Orthodoxy 1750–1850* (London: Croom Helm, 1986), 44; Hogarth, 'Joseph Townend and the Manchester Infirmary', 92.

137 Lawrence, *Medicine in the Making of Modern Britain*, 93; Hogarth, 'Joseph Townend and the Manchester Infirmary'; Riley, *Sick Not Dead*, ch. 3. Anne Digby, *Making a Medical Living: Doctors and Patients in the English Market for Medicine, 1720–1911* (Cambridge: Cambridge University Press, 1994), 86–97.

138 David M. Turner, *Disability in Eighteenth-Century England: Imagining Physical Impairment* (New York and London: Routledge, 2012), ch. 2; Roger Cooter, 'The Disabled Body' in Roger Cooter and John Pickstone (ed.), *Companion to Medicine in the Twentieth Century* (London and New York: Routledge, 2003), 367–84.

139 Riley, *Sick Not Dead*, 75.

140 For example, Durham University Library, Earl Grey Pamphlets Collection (1863), Robert Wilson, *The Coal Miners of Durham and Northumberland: their Habits and Diseases: A Paper Read before the British Association for the Advancement of Science, at Newcastle 1st of September, 1863*.

141 James Black, 'Lectures on Public Hygiene and Medical Police. Delivered at the Manchester Royal School of Medicine and Surgery, Summer Session 1844', *Provincial Medical Surgery Journal*, 28 August 1844, 555; John Benson, *British Coalminers in the Nineteenth Century: A Social History* (Dublin: Gill and Macmillan, 1980), 4.

142 Turner, *Disability in Eighteenth-Century England*, 43–4.

143 *The Miner and Workmen's Advocate*, 24, 15 August 1863; Benson, *British Coalminers*, 181.

144 The expectation of deference from patients was so strong that it was not uncommon for doctors to insist on the expulsion of recalcitrant patients from hospital: Smith, *People's Health*, 264.

145 For example, see the report on the skills of Dr Homfrey, surgeon of Tredegar Ironworks in *The Miner and Workman's Advocate. A Publication Devoted to the Interests of the Working Classes of the United Kingdom*, no. 97, 7 January 1865.

146 Riley, *Sick Not Dead*, 107, 120.

147 Kentish, *Essay on Burns*, 133–34.

148 Excerpts from 'Medical Facts and Observations, Volume the Second', *The Gentleman's Magazine*, 64:5, May 1794, 448; Turner, *Disability in Eighteenth-Century England*, 56–7.

149 David M. Turner, 'Disability and Prosthetics in Late Eighteenth- and Early

Nineteenth-Century England', in Mark Jackson (ed.), *The Routledge History of Disease* (London: Routledge, 2017), 301–19.
150 PP 1842 (381), 664.
151 John Kirkup, *A History of Limb Amputation* (London: Springer, 2007).
152 'Death Under Chloroform', *The Merthyr Telegraph*, 13 September 1862.
153 Digby, *Evolution of British General Practice*, 211.
154 *The Cambrian*, Friday 1 August 1835. On the broader context of medical negligence see Kim Price, *Medical Negligence in Victorian Britain: the Crisis of Care under the English Poor Law* (London: Bloomsbury Academic, 2015).
155 'Action Against a Surgeon for Improper Treatment', *Paisley and Renfrewshire Advertiser*, 30 March 1861.
156 *The Cambrian*, 1 August 1835.
157 *The Miner and Workmen's Advocate*, no. 17, 27 June 1863; ibid., no. 20, 18 July 1863.
158 A. D. Morris, 'Two Colliery Doctors: the Brothers Armstrong of Treorchy', in Cule (ed.), *Wales and Medicine*, 209.
159 Benson, *British Coalminers*, 181.
160 National Association of Coal, Lime, and Iron-Stone Miners, *Transactions and Results of the National Association of Coal, Lime and Iron-Stone Miners of Great Britain, Held at Leeds, November 9, 10, 11, 12, 13 and 14 1863* (London: Longman et al., 1864), 44.
161 'South Dunraven Colliery Treherbert. Notice to Terminate Contracts', *The Cardiff Times*, 16 October 1886.
162 PP 1842 (380), 185.
163 Borsay, *Disability and Social Policy*, 49–61; Julie Anderson, *War, Disability and Rehabilitation in Britain: 'Soul of a Nation'* (Manchester: Manchester University Press, 2011).
164 Benson, *British Coalminers*, ch. 7.

3

DISABILITY AND WELFARE

Writing to *The Times* in the aftermath of the Gethin Colliery explosion of 1865, Dr W. Wadham wrote movingly on behalf of the victims of the tragedy – 'for those who are dead, for those who linger in their agony' and 'for the widows and orphans of the first, and the aged and little ones depending for their daily bread upon the now no longer available labour of the latter'. These were people, he wrote, who deserved not philanthropy, but better measures to ensure safety in coal mines. Victims of disasters required no 'alms', he wrote, 'feeling that by mutual assistance, and that *heroic self-dependence* so happily distinguishing the mining population of Wales, they will know how to encounter and overcome the pecuniary distresses accompanying a catastrophe which would render those less strong of heart the recipients of public charity'.[1] Just over a decade later, a similar view was presented in the *Glasgow Herald* of Scottish coal workers. 'All who know the decent working people of this part of the kingdom', the article contended, 'are aware how reluctant they are, as a rule, to become … the recipients of charity.' It extolled the virtues of the careful, self-sufficient miner whose infirmity was supported by payments from the 'box' of his friendly society, into which he had contributed 'when able to work'. Such a man was 'regarded as a gentleman compared with one who has had to go on the parish'.[2] Contrasting with the stereotype of miners as feckless, the image of the self-sufficient coal worker, managing his own welfare needs or supported by mutual aid in his community, was a powerful ideal and motivated thousands of miners to join friendly societies, insurance schemes, pit clubs and trade unions as the nineteenth century progressed.[3]

Disability has long been at the heart of discussions and debates about welfare. It has been used as a yardstick by policymakers, charities and self-help organisations to determine not only who *needs* support and assistance, but also who *deserves* it.[4] Welfare systems played a critical role in defining 'disability' during our period and also imposed a series of responsibilities and moral strictures on those who sought support. At the same time, those claiming relief had their own ideas about the help to which they felt entitled, and these expectations also shaped their experiences. Although material support or assistance was often crucial in maintaining the economic well-being of individuals and their families, welfare, as the examples above show, was an emotional issue that evoked feelings of pity, pride, dependency, hostility and responsibility. These feelings, as much as the practicalities of financial or medical provision, shaped experiences of welfare in the past, as they continue to do so today.

As the previous chapter suggested, coalminers, like other nineteenth-century workers, were enmeshed in a 'mixed economy' of welfare that comprised many sources of assistance – from family support to self-help initiatives. It also included public charity and support from the state through the Poor Law, despite the image of the heroically 'self-dependant' worker commonly propagated by the labour movement. Due to the high rate of non-fatal accidents and industrial diseases in the sector, mining developed its own distinct ecosystem of welfare, designed to meet the specific needs of disabled coalminers. The sick clubs established by paternalist coal owners, explored in the previous chapter, were part of this, but so too were initiatives devised and promoted by miners themselves. As we shall see, the friendly societies lauded by the author of the article in the *Glasgow Herald* were not always welcoming of coalminers, particularly at the beginning of our period. Frustration with the ability of funds to cater for the needs of the long-term disabled led to new initiatives to form distinctive, occupationally focused forms of self-help, of which the most notable was the permanent relief fund movement founded in the wake of the Hartley Colliery disaster of 1862.[5]

Focusing on the general (non-medical) support available to sick, injured and impaired mineworkers, and the social and cultural principles that underpinned it, this chapter explores where disabled miners and their families stood within the matrix of welfare expectations and provisions, and how this affected their ability to secure assistance in times of need. Where did responsibility for the care and support of ill and injured mineworkers reside and what type of assistance was given? What, moreover, were the expectations of welfare claimants and providers in coalfield communities? And how did these expectations shape the experiences of disabled miners? To answer these questions, this chapter examines mineworkers' experiences of the different strands – domestic, public and voluntary – that constituted the patchwork of care and assistance available to them. While the chapter examines each of these dimensions of welfare in turn, it is important to recognise their interdependence. Individuals often drew on different forms of support simultaneously. Sixty-three-year-old former hewer Robert Young, for example, revealed the delicate web of welfare supporting him to the Children's Employment Commission in the early 1840s. No longer able to work at mining after losing his leg at a Scottish coal works seven years previously, Young was getting by with a patchwork of assistance. Mr Deans, the owner of nearby Penston Colliery in East Lothian, allowed him free housing and coal. On top of this, however, Young also relied on small weekly poor relief payments from his local parish and the occasional shilling he and his wife managed to earn 'nursing the bairns of the wives who work below'.[6] Other disabled mineworkers similarly drew on various combinations of help from different sources of assistance, as one alone was rarely ever sufficient to make ends meet. This chapter examines how mineworkers utilised these differing support mechanisms and brought them together to fashion a successful survival strategy for themselves and their families.

Domestic sites and sources of welfare

As we saw in the previous chapter, for most sick or injured miners, the primary location of care was the home. The home also occupied a central position in nineteenth-century welfare practices. Despite a growth in institutions aimed at tackling poverty, Britons in need were most often helped in domestic, not institutional, settings during the period. Eighteenth- and nineteenth-century Britons were under a moral obligation to support any incapacitated members of their families. Parents, children and even grandparents were expected to help relatives if they were in a position to do so, and most did.[7] Indeed, of all working-class communities, those centred on mining were thought by some Victorians to have had a particularly good record in this regard. In the mid-nineteenth century, one keen observer, at least, noted that 'Old pitmen' in north-east England were 'very generally supported in whole or in part by their grown-up families – a filial duty, the performance of which is far more common in the mining than in the manufacturing districts'.[8]

In principle, male heads of households were ultimately responsible for the welfare of everyone in their families, but women and children also made important contributions. As we have seen, prior to 1842, women, girls and young children had been permitted to work underground in mines, and several thousand did, especially in Scottish coal mines (around 40 per cent of the total workforce).[9] Their work provided a means of protecting households against falling into poverty and becoming reliant on poor relief. Church has estimated that about one-third of females interviewed by the 1842 Children's Employment Commission reported having dead or disabled fathers.[10] Their labour generated valuable income for the support of their families, perhaps contributing one-quarter to half of a Scottish mining household's income where all members worked.[11] Some women worked underground to support their incapacitated husbands. For example, Margaret Boxter (or Baxter), aged fifty, of Bathgate, one of the few women who worked at hewing rather than hauling coal, said that she had worked underground for twelve years, ever since her husband 'failed in his breath' and was incapacitated from working himself.[12]

Even in areas where women were not employed underground, as in the north-east of England, sending very young children to work in coal mines was a familiar survival strategy for mining families struggling to cope with the effects of disability or other family misfortune. Ann Mills testified that she had sent her son Matthew to work underground to open and close the doors used to ventilate Blaydon Main Colliery at the tender age of six 'on account of her husband's bad breath'.[13] According to Robert Franks in his report on the east of Scotland for the 1842 Children's Employment Commission, child labour in the mines was a symptom of the inadequacies of the 'Scotch system of poor law', which forced 'youthful labour into the pits at an early age, in order to raise support for indigent parents who received either inefficient relief, or no relief at all, from the parish' – although a desire to *avoid* parochial relief may have been just as important for some.[14] Time and again, young witnesses reported that their employment underground was motivated by the necessity of contributing to the family economy. Thus William Naysmith, a twelve-year-old Scottish putter testified that he came to work underground 'as it would be more benefit to my father, who is off work with bad breath. Father, mother and five other children depend on the labour of brother and myself.'[15] Injury or chronic illness seems to have forced some miners to rush their children into the mines earlier than planned. Twelve-year-old mineworker Robert Dickson, for example, reported that 'Father would not have sent me below so soon had he not been bad in the breath.'[16] While some proprietors of coal mines testified that they were unwilling to allow very young children to work underground, they said that exceptions were made for 'children of widows, or where the families are very large'.[17]

Although the 1842 Mines and Collieries Act's prohibition of the employment of all females and boys under the age of ten was hailed as preventing a great social evil, the legislation was a mixed blessing for families of disabled mineworkers. On the one hand, in banning women's work underground, the act intended to promote a 'domestic ideology' that emphasised the role of women as carers and nurturers within their families. The law had considerable success in this area and many discharged female coal workers did indeed end up in the home, as reformers had hoped. This may have represented a welfare boon of sorts for mining families, since they no longer had to pay women from outside their households to perform domestic tasks for them, as many had done prior to the 1842 Act when hard working mining wives and daughters were simply too tired and lacked the time to 'keep house' properly.[18] For disabled miners who required care, then, the increased presence of female relatives at home after 1842 may have offered an element of emotional comfort and support not available from non-relatives.

On the other hand, the legislation robbed families of disabled mineworkers of an important source of income.[19] In 1845, for example, only about 200 Scottish pit women out of those thrown out of work by the law had new jobs.[20] Some mining families with disabled members may have fared even worse from the operation of the Mines and Collieries Act, especially if they lived in company housing at the time the law was passed. Speaking in the House of Commons in May 1843, Renfrewshire MP Patrick Stewart drew his fellow parliamentarians' attention to the possible implications of the act to the welfare of mining families. Referring to a practice common in Scotland whereby disabled miners were allowed to keep their company houses providing someone in their household continued to work for the mine, Stewart raised the possibility that disabled colliers supported entirely by mine working women and children risked becoming homeless because of the 1842 Act.[21] Despite Stewart's pessimistic prediction, injured miners were not thrown out of their homes wholesale. Mine-owner paternalism remained strong in many mining districts after 1842 and company housing continued to be retained for vulnerable members of colliery communities, such as widows.[22] Nevertheless, the dire forecasts made by opponents of industrial reform in the early 1840s were probably not very comforting for incapacitated miners and their families and may have led to greater uncertainty about their future prospects.[23]

Women in mining communities made significant contributions to their families' economic well-being both before and after the passage of the 1842 Mines Act. The law may have stopped them working underground, but it did not stop them supplementing family income in other ways. Disabled mineworker Edward Rymer's mother, for instance, went gleaning for corn at harvest time to support her family during periods of pecuniary pressure, perhaps 'augmenting' her household's resources by as much as 10 per cent in the process.[24] Many miners' wives also took in lodgers to help make ends meet.[25] In times of difficulty some women may have taken even more desperate measures, resorting to theft to support sick or injured relatives. When Esther Morgan was charged with stealing forty pounds of coal from Ynyscynon Colliery tip in May 1864, for example, she claimed in mitigation that her father was a former miner and a 'cripple', and her mother was 'afflicted with cancer'.[26]

The economic input of women and children to disabled miners' households thus remained important after 1842, even if a greater premium was placed on the earnings of sons. As accounts of those killed in mining accidents demonstrate, the economic support provided by children continued as they matured into adulthood and was often vital for families containing disabled people. When twenty-three-year-old Alex Richards died in the Gethin Colliery explosion of 1865, he left behind an old blind father and a sister who were 'supported solely by the deceased'.[27] Similarly, when a reporter for the *Cambrian* visited families of victims of the Abercarn disaster in 1878, he met a mother who had lost her twenty-one-year-old breadwinner son and whose husband was 'suffering from asthma and unable to work'.[28] The loss of young people in colliery disasters was especially poignant in cases such as these where the support of sick, disabled or elderly family members was also at stake.

Despite the moral and legal pressure on relatives to support their own, people in coalfield communities could not always count on help from their families. The high mobility of coalminers, especially in north-east England where for much of our period pitmen were employed on a short-term basis via the annual bond, created a situation where wives and children might be deserted as men sought work in new collieries. In 1840 the Poor Law authorities in Easington (County Durham) complained that much of their time was spent pursuing 'fugitive husbands'.[29] Furthermore, in his 1842 report on the Scottish mining settlement of Tranent, S. Scott Alison claimed it was 'quite common for collier lads' in the area to abandon their 'helpless parents' without warning. He cited as an example a case he knew of 'an old, infirm, dying collier' whose 'son deserted him' one night. In the absence of filial support, the man was served an eviction notice.[30] Newspaper reports also documented cases of neglect. William, David and John Howells, adult colliers, received a court summons in December 1886 for failing to contribute towards the maintenance of their seventy-year-old father, 'a cripple', who as a consequence had become chargeable to the Merthyr Tydfil Poor Law Union.[31]

Public welfare

As these examples show, the support of families was considered an important bulwark against a person becoming a 'burden' on public welfare. The extent to which coalminers and their families drew on the support of the state – in the form of the Poor Law – is a matter of some debate. Both at the time and since, coalfield communities have often been characterised as self-sufficient entities.

Sick or injured miners, according to this characterisation, got by on their own or with the assistance of their families, friends or their employers and rarely had to resort to publicly funded welfare. In the early 1840s, John Wetherell Hays, Clerk of the Durham Poor Law Union, reported that there were few applications for poor relief or medical support from miners in his district since 'they have their medical aid in case of accidents at the expense of the colliery'.[32] Charles Forrest, surgeon at Hirwain Iron Works, in south Wales, however, offered another explanation for the seemingly low pauper rates reported for miners in British coalfield communities. Under the Poor Law, both before and after its reform in 1834, entitlement to publicly funded welfare was usually dependent on a person's legal place of settlement, which was often their place of birth. As Forrest noted, this situation could pose serious problems for migrant mineworkers in the coalfields, particularly during times of incapacity. 'The bulk of the population not having legal settlements here', he observed, 'are of course removed to their own parishes whenever they become incapable by loss of limb or otherwise from following their work. The amount, therefore, of disabled men actually living here is very trifling indeed.'[33] Forrest's testimony suggests a reality greatly at odds with images of mining communities in which everyone – including paternalist employers, miners and local welfare officials – all did their bit (even if reluctantly) to support disabled mineworkers. Instead of receiving help and kindness in the places they fell ill or were hurt, some sick and injured miners were simply turned away and told to look elsewhere for assistance. Clearly, not all coalfield communities were as willing to provide for the welfare needs of disabled people as some popular representations would have us believe.

Despite occasional reports of disabled miners being callously 'removed' from mining areas, however, coalfield settlements did develop ways and cultures of accommodating disability that meant most permanently injured mineworkers were probably able to stay where they were at the time of injury if they wanted to. As the coal industry expanded during the eighteenth and early nineteenth centuries an ethos of employer paternalism evolved with it – albeit one that did not provide an absolute guarantee of assistance after accidents. This was especially true for north-east England. During the early decades of the nineteenth century coal owners in the region mooted a scheme to help support the 'lame' or 'superannuated' among their workforce, funded by deductions from wages and a levy on the sale of coal. Although the scheme failed thanks to worker mistrust of their employers' motives, by the 1820s a form of sick pay known as 'smart money' paid by employers to men certified incapable of work after accidents was becoming commonplace.[34] The term echoed that used to describe payments to disabled naval personnel from the

Chatham Chest, a mutual scheme established in 1588 jointly funded by the state and from deductions from seamen's wages. Describing naval 'smart money', Daniel Defoe wrote that it was 'honourable' that a 'Poor Man who Loses his Limbs (which are his Estate)' in the service of his country, and being thereby deprived of the ability to work, should not be 'suffered to Beg or Starve'.[35] Adopting the term 'smart money' reflected the importance of mining as an industry crucial to national prosperity, while also reflecting employers' moral obligation to reward workers for their loyalty and sacrifice. With all its military associations, moreover, use of the term further shored up the image of miners as 'soldiers of industry' that was also becoming popular as Britain industrialised. While miners' 'smart money' originated in north-east England and was usually funded entirely by employers, by the mid-nineteenth century coal companies in other parts of Britain were also making similar kinds of payments to 'invalids' in their employ.[36]

While undoubtedly significant, voluntary support from employers, family and community was not always sufficient. Sometimes miners and their families had to draw on public welfare and this remained an important and frequently used safety net for them throughout the nineteenth century, despite the risk of removal under the Poor Law.[37] John Leifchild challenged the characterisation of mining communities as proudly self-sufficient and averse to poor relief when he alleged, in the 1840s, that pitmen in north-east England had opposed their employers' earlier proposed welfare scheme there partly out of fear that it might 'deprive them of a claim to the poor rates'.[38] Poor relief certainly sustained many miners. Parochial records for Wales and England reveal numerous examples of mineworkers who claimed long-term relief in the wake of serious injuries, such as Richard Thomas of Tonyrefail, Glamorgan. Starting in January 1830, Thomas received poor relief for at least a year (and possibly longer) following an 'accident in a colliery' and his inability to work.[39] In the coalfield settlement of Houghton-le-Spring in County Durham, figures for 1837–41 indicate that around half of those on relief were from mining backgrounds.[40] Furthermore, although employers might provide some medical support after accidents, this rarely covered illness or injury that occurred outside the workplace. In times of ordinary sickness, then, miners and their families who lacked other resources may have been forced to turn to parochial relief. As a witness to the 1844 Parliamentary Select Committee on Medical Poor Relief put it, in cases of 'protracted sickness' miners and their families in the north-east frequently 'f[e]ll upon the parish surgeon'.[41]

The history of public welfare in late eighteenth- and nineteenth-century England and Wales has commonly been cast in terms of hardening attitudes towards the poor, represented above all by the passage of the Poor Law

Amendment Act of 1834 (or 'New' Poor Law) that replaced laws that had operated since the Elizabethan period.[42] Under the Elizabethan 'old' Poor Law, welfare provision was highly localised and often varied quite considerably from parish to parish. With the Poor Law Amendment Act of 1834, policymakers attempted to centralise the welfare system and reduce costs by making it more restrictive. Reformers wanted to create a system that was so unappealing to the poor that only the truly destitute would apply for relief. To ensure this, the new law was based on the twin principles of 'less eligibility' and the workhouse test. This meant that relief given to 'paupers' was supposed to be at a level substantially less than the worst paid labourers could earn. Furthermore, relief, especially to those considered capable of work, was only to be provided within the confines of a dreary and strictly regimented workhouse where inmates were required to work for their support. Under the new Poor Law in England and Wales, disability was regarded as quantifiable to some extent and impaired paupers were subject to classification as either 'partial' or 'totally' disabled. 'Partial' disability, which included general infirmity, was not considered a barrier to any labour whatsoever and within workhouses both the 'partially disabled' and the able-bodied were expected to undertake appropriate work 'suitable to their condition'.[43]

The new Poor Law was not uniformly implemented, despite the best efforts of reformers. For some Poor Law unions, such as Easington in the north east coalfield, building a workhouse was a 'grievous additional expense' and was resisted for several years. Overall far fewer workhouses were built than originally envisaged.[44] Indeed, poor relief continued to be distributed for the most part *outside* institutional settings.[45] As late as 1874, fewer than one in five paupers (15 per cent) in England and Wales were receiving support in workhouses.[46] Of those receiving aid outside institutional settings, moreover, a significant proportion were disabled people. Figures for the early 1870s reveal that around half of those in receipt of 'outdoor' relief (not including 'Insane Paupers and Vagrants') were aged or disabled persons and their dependent children.[47] Such relief could consist of cash or 'in kind' payments, or a combination of both. In kind payments included food, fuel, medicines, nursing attendance, clothing and rental subsidies.[48]

Nevertheless, the threat of institutionalisation shaped attitudes towards public welfare in significant ways. Like the 'respectable' poor more generally, miners viewed workhouses with disdain and tried to avoid them if possible. In his evidence to the 1842 Children's Employment Commission, the Reverend Mr Williams, a curate from Merthyr Tydfil, declared that there was a strong 'dislike' of the new Poor Law among the mine and metal workers in the town and that were a workhouse to be built 'prejudice is so strong that the people

... would pull it down'.[49] Similarly, in his evidence to the Select Committee on Mines in the mid-1860s, former miner John Normansell stated that 'we [miners] do not like the very name, to say nothing of the sight, of the union workhouse'. Though, as we saw in the previous chapter, some miners were able to benefit from medical care provided in workhouses, this antipathy towards the workhouse and the stigma associated with poor relief prompted many miners to join friendly societies or make other provisions for themselves.[50]

Significant regional differences in welfare provision persisted throughout the nineteenth century, both before and after 1834. The most glaring concerns Scotland, which had its own Poor Law system, different from that in operation in England and Wales. Prior to 1845, when the Scottish welfare system also underwent reform, the poor in Scotland had no absolute legal right to public relief, but were dependent on the goodwill of local men of standing. In 1842, S. Scott Alison noted that it was unlikely that a miner in Tranent would get poor relief 'unless he is very ill indeed'. So 'urgent' was their 'distress' that many 'even in this frail state go out to the colliery, and do a little work', and Alison reported that he had known 'several people so situated suffer serious injury in consequence'. Indigent miners and other poor persons, moreover, were only allowed to apply for relief twice a year. This meant they could not apply when need arose, but had to wait for a designated application period to come round.[51] Although some parishes did allow 'occasional supplies' for those who fell into 'accidental distress', the welfare system observed by Alison in Tranent was slow to respond to the needs of the poor and '[g]reat privation' was 'sometimes experienced by deserving people' because of the time they had to wait for their applications to be 'attended to'.[52]

In 1845, this situation changed and the poor in Scotland became entitled to welfare providing they were considered incapable of work.[53] Whereas the Poor Law operating in England and Wales provided relief for both able-bodied and disabled paupers, in Scotland disability – 'whole' or 'partial' – was made the chief criterion for support. Defining a claimant as 'disabled' might be a contested process. Poor Law records in Scotland reveal a number of cases where the presence of impairment was not deemed sufficient to grant relief. Hugh McKay, a miner, made an application for assistance from the parish of Carluke in Lanarkshire on behalf of his eldest son who was subject to epileptic fits, only to be refused on the grounds that his son was considered 'able-bodied'.[54] In Glasgow several years later it was ruled that a young man aged 26 was not eligible for parochial relief since he was 'physically able to earn eleven shillings per week', although he was 'to some extent mentally incapacitated'.[55] Definitions of both disability and destitution were elastic and could depend on local economic circumstances that might affect an injured or impaired person's ability to earn a living. 'Want of light remunerative employment is one great cause of pauperism,' noted a report on the Poor Law administration in Scotland written in the early 1870s, and it was 'not uncommon to give relief to persons who are *not disabled for some kind of work*, if it were available'.[56]

When Parliament reformed the Scottish public welfare system in 1845, it did not insist, as it had done for England and Wales, that workhouses be built or require the institutionalisation of paupers. The new law left much of the old system intact. Parishes continued to be the administrative unit through which relief was distributed. Paupers were entitled to medical relief and it was the duty of local Poor Law officials to make sure they got this. Parishes could, if they wanted to, provide this within a poorhouse, but they were not compelled to do so. Moreover, in sharp contrast to the workhouses envisaged by policy-makers for England and Wales, Scottish poorhouses were not intended – in theory at least – to serve as deterrents to pauperism and conditions within them were not set according to the principle of 'less eligibility'. Instead, poorhouses were to serve as refuges or shelters for those 'poor persons who from weakness, fatuity of mind, or by reason of intemperance, dissipation or improvidence, are unable or unfit to take charge of their own affairs'. The official emphasis was on providing them with a 'place of security, and seclusion from the world' where their 'necessity … to live by alms' could be tested, rather than punishment or deterrence.[57]

Nevertheless, in practice poor relief in Scotland was, according to Hutchison, 'administered in such a way as to make it an unattractive solution and to encourage claimants or their relatives to seek alternative remedies to poverty'. There may have been no formal workhouse test in Scotland, but, as the nineteenth century progressed, poorhouses also became more regimented, austere and disciplinary places and worked to discourage all but the totally destitute from seeking relief within their walls.[58] By 1868, Scotland had sixty-six poorhouses, which, on average, accommodated around 8,000 to 9,000 paupers annually. Demand for places in them, however, rarely outstripped supply. In 1868, for instance, Scottish poorhouses may have housed thousands of paupers, but, with capacity for approximately 12,000 inmates, they still had plenty of space for several thousand more.[59]

In some cases, such as that of Nicholas Winn, 'a drawer or coal-pit worker … in delicate health', those offered relief by Poor Law officials in Scotland were sometimes given a choice as to whether they were institutionalised. The medical officer who attended Winn in the late 1850s offered him a place in a poorhouse, where it was thought the mineworker could be better cared for than in his current lodgings. Winn, however, refused the offer and was instead provided for by the Parish of Old Monkland outside the poorhouse in private accommodation.[60] Yet inspectors in other parishes might take a different approach. In Dalziel, North Lanarkshire, a collier, Thomas Campbell, 'partially disabled by asthma', was granted relief in the poorhouse but after refusing to go he found his support denied. Several other disabled colliers in that parish found themselves in similar circumstances after refusing to enter the poorhouse.[61]

Nevertheless, as in England and Wales, most paupers in Scotland were provided for outside the walls of publicly funded welfare institutions through a similar combination of monetary assistance and 'in kind' benefits, including 'meat and clothes' or assistance in attending hospital.[62] The notes appended to applications for relief from parochial officers provide important insights into the causes of disablement in Scotland's coalfield communities and the experiences of those miners who received relief. To take one mid-nineteenth-century example, the records of applications for relief in the parish of Carluke near Glasgow reveal a variety of physical and mental impairments that prevented mineworkers from undertaking their usual work. While many claimed relief after accidents, or as a result of the incapacitating lung diseases that were endemic in many Scottish coal mines, some mineworkers became 'disabled' thanks to other events. For instance, James Watson, a married man aged forty-three, who made an application for relief on 18 July 1856, was categorised as 'wholly' disabled due to his 'labouring under some melancholic delusion as to his future state and quite unfit to follow his employment'. His application for relief was referred to committee, where, upon receipt of medical evidence, he was admitted to the lunatic asylum in Glasgow.[63] James Fyfe, a fourteen-year-old drawer at a coal pit, applied to relief on 12 May 1858 because he had been 'severely bit by a dog and is unable to work'.[64]

Being classed as 'wholly' disabled did not necessarily mean a person was considered permanently impaired. George Moffat was declared 'wholly disabled' due to an 'attack of Bronchitis', but was 'expected soon to be able again for his usual employment'.[65] Indeed, the records often made distinctions between those declared 'wholly' disabled and those who were 'wholly at present' incapacitated from working, indicating an understanding of disability as a fluid, rather than permanent, state.[66] Showing concern about the future liabilities of the parish, many notes commented on claimants' likelihood of recovery. Thomas Kerr, whose application for relief was received in December 1859, was classed as 'wholly disabled' due to an inflammation of the internal ear, 'his recovery very doubtful'. Similarly, John McKendrick applied in the same month with a broken back, which the inspector feared was 'so serious that he will never recover'.[67] The administrative records of poor relief from Carluke furthermore reveal that the authorities were familiar with the diseases of mineworkers and this allowed them to make subtle distinctions between types of impairment and their likelihood for improvement. Thus, men temporarily impaired through attacks of bronchitis were distinguished from those such as Samuel Donaldson, a thirty-one-year-old miner, who was described as 'one of those cases of miners becoming useless in the breathing apparatus from this employment before they reach the prime of life'. Though classed as wholly disabled 'at present', the notes did not indicate much faith in his likelihood of recovery.[68] Just like their counterparts in England, whose processing of disabled applicants has been analysed by Steven King, the parochial authorities in Scottish mining communities made distinctions and judgements that indicated a nuanced understanding of different impairments and their effects on individuals' lives.[69]

Overall, despite the endeavours of national policymakers and administrators to centralise poor relief, local Guardians of the Poor still exercised a lot of discretion during the nineteenth century. The overwhelming and persistent reliance on outdoor relief in many places indicates the autonomy of local officials in determining how to tackle poverty in their communities. Although large-scale disasters, economic downturns and strikes might place the Poor Law in mining areas under considerable strain and lead to greater scrutiny of claimants, under other circumstances the authorities might respond 'flexibly and imaginatively' to the financial problems faced by sick or injured mineworkers.[70] While Poor Law officials expected mineworkers to plan for their own welfare needs via voluntary schemes such as workplace accident funds or joining benefit societies, they provided support for those who were unable to maintain their payments to friendly societies or whose benefits fell short of providing enough to live on. Under the old Poor Law, overseers made ad hoc payments to colliers and others to subsidise their subscriptions to self-help organisations, such as the one-off payment of 5s. to Thomas Llewellin in Miskin Hamlet near Llantrissant in south Wales in July 1817 'for his Benefit Club'.[71]

After 1834, guardians in English coalfields also seem to have used their autonomy to favour injured miners for outdoor relief, even though the preference of national policymakers at the time was for greater use of workhouses.[72] In the aftermath of colliery accidents, guardians were also known to disregard the friendly society benefits paid to miners when deciding what allowances to give them, despite requests from London that they should take these into account. In ignoring these national recommendations, guardians wanted to reward miners who had been prudent enough to make provisions for sickness or injury and reinforce the message that those most worthy of poor relief were those who had done everything in their power to stave off penury.[73]

Voluntary welfare

Important as the state was in times of hardship, voluntary and self-help schemes were arguably more important in providing relief for those who became sick or injured in the workplace.[74] Voluntary welfare provision was generally either reactive or proactive in nature. Reactive measures were *responses* to welfare needs that had already arisen. As we have seen, employer paternalism was an important part of reactive welfare provision for injured miners; so too were informal community responses to accidents, including ad hoc 'gatherings' to help disabled mineworkers get established in other lines of work.[75] Charitable appeals ranged from ingenious efforts on behalf of individuals, such as the raffle reportedly held in Merthyr Tydfil in April 1857 to 'enable some unfortunate youth to purchase a wooden leg', to the impressive disaster-relief funds assembled in the face of large-scale casualties that threatened to overwhelm normal sources of support.[76] Although primarily established for the benefit of the bereaved, disaster funds, such as one set up after the Abercarn explosion in 1878, also paid out money to injured survivors. Donations to these funds came from coal owners, mining communities and, thanks to the publicity given to such disasters in the press, from people living beyond Britain's coal producing regions.[77] In May 1880, for example, *The Times* reported that more than a half of the £59,000 collected for 'sufferers' of the Abercarn disaster came from a fund based in London.[78] Such reactive measures to mining accidents were heralded and encouraged as signs of Christian charity.

Alongside such community responses, nineteenth-century Britons were also expected to take *proactive* measures to mitigate the effects of future misfortune. Victorian reformers such as Samuel Smiles promoted self-help initiatives as a means of encouraging 'self-reliance and self-respect' among the working classes.[79] Accordingly miners, like other workers, were increasingly expected to take responsibility for their own future welfare needs by contributing to benefit or friendly societies. If, on the one hand, hostility to the workhouse test may have made state welfare degrading in the eyes of proud coalminers, on the other those who relied on the Poor Law sometimes faced stigma for their alleged imprudence. Layton Lowndes, chairman of the Board of Guardians in the coalmining community of Madeley, Shropshire, giving evidence to the Friendly Societies Commission of 1873, remarked of colliers that their improvidence often meant that they made inadequate payments into benefit societies to cover sickness. Miners were, he noted 'getting very high wages on short hours', but squandered them in the public house rather than using them responsibly to safeguard their families' well-being.[80] Although this stereotype of the thriftless miner has been comprehensively challenged, the impact of voluntary welfare on the experiences of sick or impaired mineworkers has received less attention.[81]

Arguably the most important vehicles for insuring against sickness and injury in the nineteenth century were friendly societies. Their core aims were to provide insurance against ill health in the form of sickness benefits to members too incapacitated for work, and payment for a respectable funeral. Emerging during the eighteenth century as a response to industrialisation, there were an estimated 7,200 friendly societies in 1801 with a membership of 648,000 – about 6 per cent of the population.[82] As the century progressed, the significance of friendly societies grew along with their membership. By 1913, an estimated one in six of the British population belonged to a friendly society.[83] Throughout the nineteenth century, mining areas had some of the highest friendly society membership rates in Britain.[84] That they did, reflects a realisation by miners and their families that ill health and injury were inevitable consequences of mining.[85] It was also an acknowledgement that bodily incapacity could cause disruption to their household economies, as the labour of key contributors was diminished. When deciding who to insure against sickness or injury, then, families tended to prioritise male breadwinners over other household members.[86]

For some mineworkers, joining a friendly society was an important rite of passage marking their first steps towards responsible manhood. Recalling his youth at the end of the nineteenth century, Edmund Stonelake wrote that he had been encouraged by his mother to join a friendly society as soon as he began to earn a 'man's wage', in order to maintain his 'independence' in the face of sickness or accident. Stonelake took the ethos of self-help to heart, noting that his favourite reading at the time was Ralph Waldo Emerson's essay, *Self-Reliance* (1841).[87] Friendly societies did not simply provide welfare relief, but also cultivated an ideal of working-class manliness based on co-operation, independence and responsibility.[88]

Friendly societies took many shapes and forms. They could be broad organisations whose membership was drawn from many occupations, or confined to a single occupation. Additionally, they could be compulsory, entirely voluntary, national or local in nature. In the mining areas of west Scotland, some friendly societies reflected sectarian loyalties, with separate Catholic and Orange Order benefit organisations emerging by the 1860s and 1870s.[89] Miners joined every type of society, though they often faced hostility from general organisations that recruited from a broad spectrum of workers. Aware of the greater bodily risks that mining involved, and fearful that mineworkers would place an excessive burden on their funds, some societies banned colliers from joining altogether. The Star Friendly Society, formed at the Three Cranes in Pontypool in 1831, for example, forbade 'any Collier or other underground workman, or persons working in Lead or Copper works' from joining and warned that 'if any member shall go to work in any underground work, and receive any hurt there, he shall receive no benefit from this society for such hurt'.[90] Given this hostility, mineworkers often joined societies established specifically to cater to their needs, or formed societies of their own. Examples of such occupational societies include the United Colliers and Miners' Society based in Pontypool, established in 1831, the Beamish Colliery Friendly Society in County Durham (1835) and the Carfin Colliery Friendly Society of Lanarkshire.[91] Some, like the Rickleton and Harraton Outside Collieries' Relief Fund, established in 1833, were heavily influenced by mine owners who appointed officers and managed the funds and differed little from compulsory pit clubs; others were run by miners themselves.[92]

Local schemes had advantages and disadvantages. Of the advantages, local societies were generally quite small so members had a good chance of knowing each other personally. For disabled miners, these personal connections may have made benefit applications easier and increased the likelihood of a successful claim as familiarity and trust were important factors in welfare provision more generally in late eighteenth- and early nineteenth-century Britain.[93] For the societies, this intimate knowledge of their membership meant they had an appreciation of the personal circumstances of claimants, so were in a good position to guard against fraudulent applications.[94] However, local societies also presented a number of problems. Their small membership base meant that their benefit funds were not very large and there was always the risk that they may become insolvent, especially when mining disasters struck and multiple claims were submitted at the same time.[95] Furthermore, as social commentator Frederick Morton Eden noted in 1801, the localism of small societies could restrict the geographical mobility of insured workers, as many imposed residential requirements of one sort or another on members.[96] To be sure, some local societies did allow members to move away and still retain their right to benefits, but these societies were often ill prepared to examine, monitor or pay claimants in far off places. They were simply too small and lacked the personnel and structure to effectively administer benefits over long distances. Consequently, members who lived beyond a local society's home territory often faced additional hurdles if they wanted to claim benefits, such as having to provide legal documentation to prove their sickness or give notice of their movements.[97] For members living a long way from their societies, keeping up regular membership payments to maintain their cover could also prove challenging. Given that British miners were a geographically mobile people, local societies were not ideal vehicles for providing them with the kind of sickness and incapacity cover their itinerant and dangerous lives necessitated.

During the course of the nineteenth century, friendly societies became much better at meeting the needs of the highly mobile industrial workforce, thanks largely to the growth of the so-called affiliated orders.[98] These were friendly societies, such as the Ancient Order of Foresters or the Order of Oddfellows, that were national in scope and had large inter-occupational memberships. Due to their wide membership base and national reach, these organisations had larger funds than local societies and were generally based on firmer actuarial foundations. Proudly independent organisations, they were also more reliable and allowed members to move around the country in search of work and still maintain their accident and sickness cover. They were also more likely to allow colliers to join their ranks than other societies, and for many miners the affiliated orders offered a valuable alternative, or supplementary source of support in times of sickness, to that provided by compulsory pit clubs.[99] These national societies became increasingly important to the mixed economy of welfare in coalfield settlements as the nineteenth century progressed. By 1863, their ubiquity in the colliery communities of north-east England was so great in fact that Dr Robert Wilson felt confident enough to state that '[i]n all [pitmen's cottages] but the Skip Jack, you find the emblem of either the Foresters', Oddfellows', or Free Gardeners' Benefit Society.'[100]

Although the affiliated orders may have been more accommodating to mineworkers, however, concerns about coalminers' susceptibility to accident and illness persisted and led to calls for differential treatment. Statistics on illness cover provided by friendly societies returned to Parliament in 1853 showed that at age twenty the sickness rates amongst colliers were 36.44 per 100, compared to an average for England and Wales of 26.62, which rose to 38.96 by the age of forty-five and 50.52 by the age of sixty. While many friendly societies sought to keep their costs down by excluding new members from joining above the age of forty-five, the higher levels of sickness among younger coalminers relative to other occupations led to increasing calls for miners to pay more for friendly society benefits.[101] A witness to a parliamentary select committee in 1849 suggested that 'workmen in coal mines and iron mines' ought to pay contributions 'probably twice as much as the agricultural population should pay'.[102] After debating whether 'under ground workmen should be admitted on the same footing as other work men', the Bute lodge of Oddfellows in Llantrissant agreed in 1873 to admit miners on condition that their benefits would be a shilling per week less than those paid to 'any other workman'.[103]

By this time, it was becoming common for some societies to charge 'miners and colliers' a 'special rate as to sickness'.¹⁰⁴ As F. G. P. Neison wrote in an article, 'The Influence of Occupation on Health' published in *The Foresters' Miscellany* in 1874, the particular liability to injury or sickness of different occupations made a monolithic approach to sickness benefits impractical, since injuries that would 'suffice to disable sawyers, colliers, or miners, would have but little effect on those following quiet and sedentary occupations'. Therefore, he argued, societies should adapt their rules to the occupational composition of their members.¹⁰⁵ Although members of the Foresters opposed plans to charge higher membership rates for miners in the late nineteenth century, arguing that the average numbers of sick days claimed by coalminers was in fact not significantly higher than for other workers, symbolically at least the greater exposure to risk faced by colliery workers bolstered calls for differential treatment.¹⁰⁶

Regardless of their respective strengths and weaknesses, all friendly societies shared certain similar characteristics. As Simon Cordery notes, in general, most societies 'had age, health, occupational, and moral prerequisites to membership'.¹⁰⁷ Many required potential members to be free of ill health or impairment at the time of joining. Good health and good moral character were inextricably linked in the membership requirements of many societies.¹⁰⁸ For instance, the rules of the Kilmarnock Coal Cutters Friendly Society, established in 1834, stated that 'all persons admitted ... must be of a good moral character, free of bruise or any bodily infirmity and capable to gain a livelihood for himself and family'. It also encouraged members to report others who might be concealing a 'bodily infirmity' when they joined.¹⁰⁹

Many societies similarly used moral criteria to define injuries or impairments worthy of financial assistance. The Sons of the Globe in Monmouthshire declared that if any of their members 'shall be disabled by gambling on the Sabbath day he shall receive no relief from the fund for such his misfortune', and only provided relief for those 'disabled by fighting' if it could be proved that they were not the 'aggressor'. Those who might 'wilfully maim or hurt' themselves or 'feign sickness, lameness, or other acts of dissimulation' to gain relief would be expelled.¹¹⁰ The long list of exclusions set out by the St George's Friendly Society in 1820, including ill health caused by 'venereal disease, old ulcers, sores, jumping, wrestling, fighting, gaming, hunting, or any other improper conduct, or from being concerned in any riot or mob', was typical of many in using the *causes* of illness or injury, rather than its *consequences*, in determining access to benefits, using moral criteria to draw distinctions between 'good' and 'bad' sick and disabled claimants that reinforced centuries-old distinctions between 'deserving' and 'undeserving' applicants for relief.¹¹¹ Although some men balked at the puritanism of some societies – such as those who worked at Graig Colliery in Merthyr Tydfil who, as a witness to the 1842 Children's Employment Commission noted, 'cry out against the bastardy clause' – many still availed themselves of their services.[112]

To ensure moral probity, those applying for benefits were often required to mobilise the support of respected persons to verify their claims. Indeed, these same figures were also regularly called upon to examine and certify the incapacities of applicants. In the first half of the nineteenth century, examination of claimants was 'social' as well as 'medical' with prominent laypersons providing support for applications alongside doctors and surgeons.[113] As late as 1873, the Tranent Miners New Friendly Society ordered claims to be verified either by a surgeon or a 'minister of some religious denomination', suggesting that medical authority in defining sickness was not yet absolute.[114] The involvement of lay persons alongside medical experts is evident in the extensive correspondence that survives for the Glais Friendly Society, located at Llansamlet near Swansea. Although a 'surgeon' certified some claims, other applicants had their applications additionally verified by clergymen, churchwardens or representatives from their place of work. Therefore when John Morgan of Llansamlet applied for assistance when he became unable to work due to a 'sore leg' in 1844, his application was certified by a surgeon, a churchwarden, the parish curate and David Lloyd, the agent for Cymllynfell Colliery where he presumably worked.[115] Henry Rees's claim in the summer of 1857 was similarly supported by a surgeon, an independent minster and Thomas Walters, a colliery overman.[116] A good relationship with colliery managers, then, was clearly important for some mining claimants, though it was rarely an absolute requirement for a successful claim. The significance of doctors in the affairs of nineteenth-century friendly societies also varied. During the first half of the century, societies tended to employ doctors primarily in a 'policing role', screening applicants for relief where needed. In the late 1830s, for example, Hebburn Colliery Relief Society allowed its officers to consult surgeons for their opinion in cases of suspected fraud.[117] With time, however, doctors generally became more central to the business of friendly societies and they were increasingly called upon to provide members with medical care as well as validate claims. By the end of the century, doctors were used by friendly societies almost as much for their therapeutic skills as for their supposed ability to assess the cause and extent of sickness objectively.[118]

To guard against fraud and ensure that they maintained their good moral character, society officials also visited the sick and incapacitated regularly.[119] For claimants, including disabled miners whose impairments were acquired after the start of their membership, such visits may have been considered intrusive and viewed with trepidation. Patients were expected to adopt a 'sick role' in which they agreed to passively submit to medical advice and to remain confined to their homes.[120] While claiming sickness benefits, members of Wishaw Iron Works Miners' Annual Friendly Society were expected to avoid 'public places of amusement', refrain from drinking or gambling and 'shall not leave home without obtaining and handing to the President a Doctor's certificate that his going from home is for the good of his health'.[121] At the same time, however, the visitations made by society stewards could be a source of comfort, especially in small local societies where stewards were more likely to be personally acquainted with benefit recipients.[122]

Payments from friendly societies varied and for smaller societies might depend on the overall health of the fund. Some injuries merited higher payments. In south Wales, the rules of a friendly society at Risca made in 1817, for example, allowed weekly payments of between three shillings and ten shillings, depending on the financial state of the society, but specified a minimum payment of one guinea (£1 1s) if a member lost a limb and allowed additional relief for members with broken bones. This was in recognition of the higher medical costs involved in treatment.[123] Isaac Williams a collier member of the Sons of the Globe Society of Pontypool, received a one-off payment of £1 17s 3d 'all[owe]d for broken bones', together with weekly sick pay of four shillings in 1821.[124] However, most societies were not geared up to meeting the *long-term* welfare needs of disabled claimants. Some societies ceased their support completely after a period of time, placing members under pecuniary pressure. In Merthyr Tydfil in 1849 it was reported that some of the 184 men relieved by the Poor Law Union were victims of accidents in the town's collieries or iron works who had become 'burdensome to the parish' because they had 'exhausted their "sick fund" or "benefit society"' allowance.[125] Most societies reduced payments after several months. Carfin Colliery fund, for example, paid 6s a week to those 'disabled from working', which was reduced to 5s after four months and 4s after eight months. If the claimant had not recovered after a year 'he shall be placed on the superannuation list which is declared to be two shillings weekly'.[126]

There is anecdotal evidence that in some areas workers may have drawn on several benefit societies to help make ends meet. David James, chairman of the Merthyr Board of Guardians, told the *Morning Chronicle*'s reporter in 1849 of a man who earned 14s a week at work, but when certified sick earned 18s a week – 7s from each of two friendly societies he had joined, and 4s from his work's sick fund.[127] The idea that workers' sickness benefits might exceed their wages caused considerable concern for 'respectable' Victorians.[128] Indeed, some societies were so worried about this possibility that they guarded against it by expressly prohibiting members from belonging to more than one society at a time.¹²⁹ For most people, however, reliance on benefits from a society undoubtedly represented a drop in income. In the late 1830s and early 1840s, Tranent miners claiming sickness payments from a 'benefit society' were reported to receive five shillings a week, at most. Considering that colliers in the area at the time were reputedly able to earn twenty shillings a week quite comfortably, the benefit available to disabled miners can hardly be regarded as generous. At around a quarter of what a miner might expect to earn at work, receipt of such benefits would have meant a significant fall in income for mining households. They were, in short, a poor compensation for a full loss of a miner's earnings.¹³⁰

Unlike the Poor Law's distinction between 'whole' and 'partial' disability, friendly societies generally adopted a simple concept of incapacity in which a person was either able to work or not. The rules of most societies prohibited payments to sick and injured members who worked.¹³¹ There were occasional exceptions. In 1857, the Oddfellows Bute Lodge in Llantrissant agreed to pay £10 to a quarryman, Richard Williams, 'to enable him to commence some business in consequence of the loss of his hand from an accident', in lieu of weekly payments.¹³² However, for the most part, miners and other workers with long-term health problems had to choose between receiving a meagre benefit and working as best they could. Without doubt, friendly societies offered a significant safety net for miners and other nineteenth-century workers that fostered pride in self-reliance, and was free from the stigma of the Poor Law or (in many cases) the controls of employer paternalism. But for most miners in receipt of sickness benefits, societies worked in such a way as to deprive them of a means of supplementing their incomes through light work, which might relieve some of the financial pressure injury and illness placed on them and their families.

Permanent relief funds and the politics of welfare

As the nineteenth century wore on, the problem of how to effectively support the long-term welfare needs of disabled coalminers and their families attracted other solutions. Trade union accident funds, which are examined in more detail in Chapter 5, operated in similar ways to friendly societies.¹³³ In the second half of the century, life insurance companies also sensed a commercial opportunity in providing cover for workers in collieries and other dangerous occupations.¹³⁴ For example, *The Miner and Workmen's Advocate*, the self-declared 'Publication devoted to the interests of the Working Classes of the United Kingdom', recommended that its readers join the Friend in Need Life Assurance company, citing the weakness of friendly societies in providing adequate cover in cases of chronic sickness or old age. Faced with the daily threat of being 'lamed, crushed, blinded, burnt and injured in every conceivable way', readers were told to 'select a proper institution and insure at once'.[135]

However, the most significant step towards tackling the long-term needs of disabled mineworkers was taken in the aftermath of the Hartley Colliery disaster, which claimed 204 lives in January 1862. The accident led not only to an outpouring of public sympathy, but also triggered debates about how the victims of all mining accidents should be supported. In spite of raising some £85,000 in a disaster-relief fund to help the bereaved, the response to the catastrophe 'highlighted to miners of the north the injustice of a system that overprovided for families of [victims of] disasters', while not providing long-term support for those whose loved ones were killed 'singly' in quotidian accidents that never 'reached the public ear'. It led to calls for more durable systems of support so that miners permanently disabled did not have to rely on ad hoc fundraising, nor pit clubs, which 'seldom make substantial provision for more than a few months of disablement'.[136] These concerns became the catalyst for the formation of the Northumberland and Durham Miners' Permanent Relief Fund.

The fund had two main objectives: providing relief for widows and orphans of men killed in mining accidents and providing support for the long-term disabled. It was established on 7 June 1862 and within two weeks had 2,000 members from thirty collieries. By October that year its membership had risen to 7,560 from sixty-one collieries and by the end of its first year it had almost 8,000 subscribing members.[137] In 1863, surplus money from the Hartley Relief Fund was amalgamated into the Northumberland and Durham Miners' Permanent Relief Fund. By the time of its second annual report in May 1864, the fund was reported to be in good health, despite the expenditure on permanent disablement being 'far greater than had been expected'. There were nine adults and two youths receiving disablement benefits costing £205 0s 8d a year (about a quarter of its expenditure).[138] By 1868, membership had grown to 11,000 and despite initial scepticism about the likely success of the scheme, many colliery owners were now encouraging their workers to join. The proprietor of Walbottle Colliery, for example, offered men an additional three shillings a week smart money for members as an inducement for men to join the scheme.[139] By the 1870s, colliery owners were actively subscribing to the fund, contributing £20 for every £100 they paid to their workers.[140]

The permanent relief fund aimed to support the victims of serious accidents, such as where limb amputation had taken place or where men had received spinal damage 'whereby [they were] not able to work any longer'.

In an attempt to distinguish between longer-term conditions and temporary impairments, payments started after twenty-six weeks' absence from work. As Alexander Blyth, secretary of the society, reported in 1872, disabled members received eight shillings a week, paid for by membership fees of three-and-a-half pence a week. The fund differed from conventional friendly societies in that it aimed to assist those permanently disabled in mine work (as a result of accidents on the surface as well as underground), and potentially made payments in perpetuity. However, in response to the objections of some members about having to wait half a year before they received any benefits, a 'minor accident' fund was established in 1869. This promised five shillings a week to injured members during the first twenty-six weeks of incapacity. After that, if they were still unable to work, they were transferred to the permanent relief fund. To enable this transfer between schemes, only men who were already members of the permanent fund were allowed to sign up for the 'minor accident' fund. In addition, there was also a sickness fund, which varied its payments according to the age of members, although this was not well used. The Northumberland and Durham Permanent Relief Fund therefore offered a variety of benefits, providing similar services to friendly societies as well as more innovative long-term relief to the permanently disabled.[141]

Two-thirds of members, estimated Blyth, were also members of friendly societies such as the Foresters or Oddfellows, showing that the permanent relief fund was expected to *supplement* rather than *replace* other forms of support. In providing higher payments to long-term disabled members, compared to those disbursed to victims of 'minor accidents', however, it marked a distinct change from normal friendly society policy.[142] Whereas friendly societies customarily reduced their payments over time, the permanent relief fund recognised the long-term needs of its disabled members. Its principles were adopted elsewhere. Proposals for a Scottish Permanent Relief Fund in 1878 similarly recommended payments of eight shillings a week for miners disabled and 'unable to gain a livelihood in any other employment' and proposed payments at the same level for both widows and 'dependent relations of permanently disabled members' – a decision that equated permanent incapability to earn a living with death.[143] By 1880, permanent relief funds were found across English coalfields and had enrolled a fifth of the country's miners.[144]

The first disabled claimant of the Northumberland and Durham Miners' Permanent Relief Fund was Henry Baker of Backworth Colliery, who received help from 8 January 1863 to his death in March 1875.[145] By 1880, the income from the 70,663 members of the Northumberland and Durham fund was around £37,380, providing support for some 232 permanently disabled members and 1,110 'aged miners'.[146] Nevertheless, the permanent relief fund faced a number of problems and criticisms. Like other welfare schemes there were fears that it was open to fraud, and from an early stage the Northumberland and Durham fund engaged the services of medical men to enquire into 'all cases of permanent disablement' and to regularly examine 'disabled members'.[147] Administrators of the disablement fund also faced 'many and great difficulties in determining the difference between diseases arising from natural causes and diseases caused by accidents'.[148] The cost of supporting long-term 'disablement' also threatened the financial health of the fund. In 1864 the Northumberland and Durham Permanent Relief Fund passed a motion to 'grant a disabled member a sum of money wherewith to commence business' in place of paying future claims, in the hope that those capable of some work would take the opportunity to be self-sufficient.[149] Strikes, such as the one in north-east England in 1879, reduced membership payments, putting pressure on resources.[150] The large number of disabled and aged miners taking advantage of the scheme by the end of the 1870s led to calls for some funds to raise membership subscriptions.[151]

The partnership with employers – who assisted with the management and administration of the funds – and the local, rather than national, organisation of the movement, also raised concerns.[152] At its inception, the Northumberland and Durham scheme was criticised in the pages of *The British Miner and General Newsman* and *Miner and Workman's Advocate* as being merely a 'local fund', built on 'sandy foundations', where the involvement of mine owners made recipients vulnerable to losing their benefits for 'presumed insubordination'.[153] This animosity stemmed from initial talks between the Permanent Relief Fund Society and the National Association for the Relief of British Miners led by trade unionist John Towers, at which the latter had insisted that the permanent relief fund come under its auspices.[154] Towers used his *British Miner* newspaper to advocate a national, trade-union-led solution to the problem of death and disablement. Alongside paying benefits to dependents of those killed in accidents, providing 'such suitable provision as the case may require' in cases of disablement and a superannuation allowance to the aged or incapacitated, the National Association (which later became the British Miners' Benefit Association) also promised to fight for improved safety to prevent accidents from occurring in the first place, arguing that a 'radical cure' was better than 'palliative' care.[155] As Blyth noted in 1872, this was 'more a political society than a benefit society' and despite offering the appeal of independence, it was seen by supporters of the permanent relief fund as impractical since it was 'not likely to get the support of the owners of the collieries' needed to secure long-term financial success. Eventually it foundered.[156] Nevertheless, as permanent relief funds became established in other coalfields and expanded their membership, the need to form a national umbrella organisation to oversee the administration of the scheme was eventually accepted.[157]

Conclusion: disability, eligibility and welfare

The experience of welfare in the coalfields was marked by variability. Sick, injured and impaired miners could have very different experiences depending on when, where and who they asked for help. The 'mixed economy' of welfare that sustained disabled mineworkers and their families in times of need was multi-faceted. Its three main components, domestic, voluntary and public welfare, were closely inter-related and in times of distress a miner might draw on several sources of support simultaneously. Although reliance on public welfare called into question a person's respectability, it did provide an important safety net for many disabled miners.[158] Stories of workers profiting from their incapacity by combining benefits from multiple sick pay schemes to earn more than they did at work were probably apocryphal. Most injured miners struggled to get by on benefits that usually fell far short of their normal wages. That such stories circulated, however, indicates the deep suspicion of welfare claimants in nineteenth-century Britain, as well as the expectation that miners, with their relatively high pay, would make their own provisions for incapacity. This expectation was embodied in the cultural ideal of 'heroic self-dependence', of the man who strove to retain his independence in the face of misfortune. It was also reflected in idealised portrayals of miners who refused to become a 'burden' on the poor rates and chose instead to work through their impairment. Giving a lecture on 'eminent miners' to Bristol Mining School in March 1857, colliery owner Handel Cossham related the story of a Cornish tin miner, 'the Blind Miner of Bottalack', who worked in a mine in spite of his visual impairment so that he could support his 'large family' and avoid reliance on poor relief. 'This noble miner has left an example of true independence,' concluded Cossham, 'would that it was more common among pitmen.'[159]

Coalminers faced regular accusations of fecklessness such as this, but the popularity of friendly societies in the coalfields – in spite of the reluctance of some to admit miners – shows that the ideal of independence was taken to heart by many.[160] Like other welfare claimants, miners faced suspicion of fraud. To guard against this, welfare schemes – from the Poor Law to workplace sick clubs and friendly societies – put in place mechanisms to screen applicants. These mechanisms increasingly came to rely on medical surveillance as the period progressed.[161] Yet it should also be remembered that suspicion went both ways and there were many reports of coalminers distrusting the motives of welfare providers. The reluctance of Durham miners in the early nineteenth century to accept a system of accident provision that might make them ineligible for poor relief shows how miners were reluctant to subscribe to any scheme that threatened what they saw as a right to relief from parochial sources.[162] While employer paternalism in the form of medical care and sick pay played a significant role, attempts to expand this provision were seen by some as an attempt to exert greater control over workers, especially as trade unionism developed.[163] The success of the permanent relief fund movement showed the value of co-operation between workers and employers in providing viable long-term assistance to disabled miners, but not every miner welcomed it. While the movement spread across the English coalfields in the 1860s and 1870s, in south Wales it faced greater opposition, coming up against (in the view of George Campbell, secretary of the Lancashire and Cheshire Permanent Relief Society) the 'prejudices of miners against anything new'.[164]

Whereas the Poor Law and voluntary welfare provided by friendly societies were both important in addressing the problem of work-related incapacity, their approaches were different. Although inability to work was central to the notions of entitlement that animated the support mechanisms they administered, there were significant differences regarding how officials approached the question of incapacity. For friendly societies, the cause of a claimant's 'disability' or sickness was as important as its presence. This was less so in the context of the Poor Law. Friendly society rules often made moral considerations about inappropriate, and therefore ineligible, causes of incapacity. This meant it was possible for two miners with identical impairments and levels of incapacity to apply to a society for benefits and be treated in radically different ways depending on the causes of their injuries. Moreover, by prohibiting benefit recipients from working, many friendly societies effectively forced disabled miners to act out a 'sick role', in which they had to forgo certain ordinary social activities. In contrast, in Poor Law contexts, paupers could be recognised as sick, but still forced to work to some extent if they were capable of doing so. When it came to public welfare, enforcing the work ethic was paramount and idleness, for all but the totally incapacitated, was frowned upon.[165]

Friendly societies reinforced cultural ideals of working-class self-reliance, but in placing time restrictions on benefits, their funds were geared towards assisting with short-term sickness or incapacity. The permanent relief fund established in Northumberland and Durham in 1862 provided welfare on a different basis, allowing a more generous settlement for men left incapable of labour than that allocated to victims of 'minor accidents'. Whereas in other welfare contexts the term 'disability' was often used in loose ways, to refer to temporary as well as long-term incapacity, the permanent relief fund used 'disablement' to refer to life-changing events that affected a person's livelihood in decisive and 'permanent' ways. Its emergence therefore represented an important moment in the history of disability and welfare, recognising that those with long-term conditions needed distinctive forms of relief that were not bound by temporal limitations.

The financial cost of mining accidents was considerable. But the impact of disablement was more than simply medical or financial – it also affected the standing of miners in their communities and relations within their families. The following chapter turns its attention to the question of how social relations in coalfield communities were affected by disability.

Notes

1 *The Times*, 30 December 1865. Our emphasis.
2 'The Miners' Relief', *Glasgow Herald*, 20 March 1878.
3 John Benson, 'The Thrift of English Coal-Miners 1860–95', *Economic History Review*, 31:3 (1978), 410–18; John Benson, 'English Coal-Miners' Trade-Union Accident Funds, 1850–1900', *Economic History Review*, 28:3 (1975), 401–12; John Benson, 'Coalowners, Coalminers and Compulsion: Pit Clubs in England 1860–80', *Business History*, 44:1 (2002), 47–60; John Benson, *British Coalminers in the Nineteenth Century: A Social History* (Dublin: Gill and Macmillan, 1980), ch. 7.
4 Deborah A. Stone, *The Disabled State* (London: Macmillan, 1985).
5 John Benson, 'Coalminers, Coalowners and Collaboration: the Miners' Permanent Relief Fund Movement in England, 1860–1875', *Labour History Review*, 68:2 (2003), 181–94.
6 PP 1842 (381), *Appendix to the First Report of the Commissioners. Mines. Part 1. Reports and Evidence from the Sub-Commissioners*, 460.
7 Norman McCord, 'Aspects of the Relief of Poverty in Early 19th-Century Britain', in R. M. Hartwell et al. (eds), *The Long Debate on Poverty: Eight Essays on Industrialisation and 'the Condition of England'* (London: Institute of Economic Affairs, 1972), 94–5; M. A. Crowther, 'Family Responsibility and State Responsibility in Britain before the Welfare State', *The Historical Journal*, 25 (1982): 131–45.
8 Jules Ginswick (ed.), *Labour and the Poor in England and Wales 1849–1851: the Letters to the Morning Chronicle from the Correspondents in the Manufacturing and Mining Districts, the Towns of Liverpool and Birmingham, and the Rural Districts*, vol. 2: *Northumberland and Durham, Staffordshire, The Midlands* (London: Frank Cass, 1983), 63.
9 Angela V. John, *By the Sweat of their Brow: Women Workers at Victorian Coal Mines* (London: Routledge and Kegan Paul, 1984), 24; Robert Duncan, *The Mineworkers* (Edinburgh: Birlinn, 2005), 35, 70.

10 Roy Church, *The History of the British Coal Industry*, vol. 3: *1830–1913: Victorian Pre-Eminence* (Oxford: Clarendon Press, 1986), 191.
11 PP 1842 (008), *Sanitary Enquiry Scotland: Reports on the Sanitary Condition of the Labouring Population in Scotland, in Consequence of an Inquiry Directed to be Made by the Poor Law Commissioners*, 85 and 94.
12 PP 1842 (380), *Commission for Inquiring into the Employment and Condition of Children in Mines and Manufactories. First Report of the Commissioners*, 19; PP 1842 (381), *Appendix to First Report of the Commissioners, Part 1*, 475.
13 PP 1842 (381), 621.
14 Ibid., 402.
15 Ibid., 452.
16 PP 1842 (381), 455.
17 PP 1842 (380), 19.
18 John, *Sweat of Their Brow*, 43; Duncan, *The Mineworkers*, 111.
19 5 & 6 Victoria Cap. XCIX, *An Act to Prohibit the Employment of Women and Girls in Mines and Collieries, to Regulate the Employment of Boys, and to Make Other Provisions Relating to Persons Working Therein*, 10 August 1842.
20 John, *Sweat of their Brow*, 53.
21 *Hansard*, HC Deb, 16 May 1843, vol. 69, cols 475–6.
22 Although this support was threatened during industrial unrest, as we shall see in Chapter 5.
23 Benson, 'English Coal-Miners' Trade-Union Accident Funds', 402; M. J. Daunton, 'Miners' Houses: South Wales and the Great Northern Coalfield, 1880–1914', *International Review of Social History*, 25:2 (1980), 143–75.
24 Edward Rymer, *The Martyrdom of the Mine* (Middlesbrough, 1898), 5; Steven King, and Geoffrey Timmins, *Making Sense of the Industrial Revolution: English Economy and Society, 1700–1850* (Manchester: Manchester University Press, 2001), 297. For more on gleaning at this time, see Peter King, 'Gleaners, Farmers and the Failure of Legal Sanctions in England 1750–1850', *Past & Present*, 125 (1989), 116–50.
25 Andrew Walker, '"Pleasurable homes"? Victorian Model Miners' Wives and the Family Wage in a South Yorkshire Colliery District', *Women's History Review*, 6 (1997), 317–36.
26 *Aberdare Times*, 14 May 1864.
27 *Merthyr Telegraph*, 23 December 1865.
28 *The Cambrian*, 27 September 1878.
29 TNA, MH12/3052, Easington Union Poor Law Correspondence, Letter from the District Poor Law Commission Representative, 13 November 1840.
30 PP 1842 (008), 101.
31 *Aberdare Times*, 11 December 1886.
32 PP 1842 (381), 153.
33 PP 1842 (382), 553.
34 Northumberland Record Office, 263/A1/1, The Committee of Coal Owners

of the Rivers Tyne and Wear, Committee Minute Books, 1805–15, Letter from William Putter to John Buddle, 3 June 1812.
35 Daniel Defoe, *Essays upon Several Subjects: Or Effectual Ways of Advancing the Interests of the Nation* (London: Thomas Ballard, 1702), 125.
36 See for instance *Statistical Compendium*, table 5.5, Govan Colliery Journals, Payments made to disabled miners, 1852–55, http://doi.org/10.5281/zenodo.183686, accessed 24 March 2017.
37 C. G. Hanson, 'Craft Unions, Welfare Benefits, and the Case for Trade Union Law Reform, 1867–75', *Economic History Review*, 28 (1975), 244.
38 PP 1842 (381), 723.
39 Glamorgan Archives, P62/31, Llantrissant Parish Overseers Accounts, 1825–37, 142–3, 152, 161–2, 172, 181. The name 'Richard Thomas' of Tonyrefail appears over the next five years, but it is not possible to ascertain whether this is the same person.
40 PP 1842 (381), 718.
41 PP 1844 (531), *Report from the Select Committee on Medical Poor Relief*, 270.
42 The literature on the old and new Poor Law is voluminous. See S. G. Checkland and E. O. A. Checkland (eds), *The Poor Law Report of 1834*. (Harmondsworth: Penguin Books, 1974); Anne Digby, *The Poor Law in Nineteenth-Century England and Wales* (London: Historical Association, 1982); Derek Fraser (ed.), *The New Poor Law in the Nineteenth Century* (London: MacMillan, 1976); J. D. Marshall, *The Old Poor Law, 1795–1834*, 2nd edn (Basingstoke: Macmillan, 1985); John Knott, *Popular Opposition to the 1834 Poor Law* (London and Sydney: Croom Helm, 1986); Steven King, *Poverty and Welfare in England 1700–1850: A Regional Perspective* (Manchester: Manchester University Press, 2000); Steven King, '"Stop This Overwhelming Torment of Destiny": Negotiating Financial Aid at Times of Sickness under the English Old Poor Law, 1800–1840', *Bulletin of the History of Medicine*, 79:2 (2005), 228–60.
43 See, for instance, TNA, MH2/1 Poor Law Commission: Rough and Classified Minute Books, 23 August to 31 December 1834; PP 1836 (595), 'Second Annual Report of the Poor Law Commissioners for England and Wales', Appendix A, 80; Gwent Archives, CSWBGP/M1, Pontypool Union Board of Guardians Minutes, 8 December 1838, 363.
44 TNA, MH12/3052, Easington Union Board of Guardians to Poor Law Commission (PLC), 14 October 1834 (received 16 October 1834).
45 Anne Borsay, *Disability and Social Policy in Britain Since 1750: A History of Exclusion* (Basingstoke: Palgrave Macmillan, 2005), 25–36; Iain Hutchison, *A History of Disability in Nineteenth-Century Scotland* (Lewiston, NY: Edwin Mellen Press, 2007), 131–70; David Ashforth, 'The Urban Poor Law', in Fraser (ed.), *The New Poor Law*, 135.
46 Fraser, 'Introduction', 18 (Table II); Kim Price, 'The Crusade Against Out-Relief: A Nudge from History', *The Lancet*, 377:9770 (19 March 2011), 988–99; Elizabeth T. Hurren, *Protesting about Pauperism: Poverty, Politics and Poor*

Relief in Late-Victorian England, 1870–1900 (London: Royal Historical Society, 2007).
47 PP 1871 [C.396], *Twenty-Third Annual Report of the Poor Law Board*, 357, 358, 377, 378.
48 Digby, *The Poor Law*, 8. For an early example from a Welsh mining community, see Glamorgan Archives, P62/5, Llantrisant parish: Vestry minute book, 1802–15, 444.
49 PP 1842 (381), 506.
50 PP 1866 (431), (431-I), 'Index to the Report from the Select Committee on Mines', 117; Digby, *The Poor Law*, 32; PP 1842 (381), 506.
51 PP 1842 (008), 102, 114.
52 Ibid.; The Statistical Accounts of Scotland 1791–1845: Blantyre, County of Lanark, Account of 1791–99, 220, http://stat-acc-scot.edina.ac.uk/link/1791-99/Lanark/Blantyre/2/220/, accessed 27 September 2016.
53 Hutchison, *A History of Disability*, 26–9.
54 Mitchell Library, Glasgow City Archives, CO1 27 97, Parish of Carluke, Record of Application for Relief 1866–77, no. 749, application for relief by Hugh McKay, 18 November 1869.
55 *Scottish Poor Law Magazine*, Vol. 5 (1871–72), 519.
56 Ibid., 106. Our emphasis.
57 'An Essay on the Poor Law of Scotland', *Scottish Poor Law Magazine*, 4 (1870–71), 139–40.
58 Hutchison, *A History of Disability*, 284, 185, 190.
59 Audrey Paterson, 'The Poor Law in Nineteenth-Century Scotland', in Fraser (ed.), *The New Poor Law*, 190.
60 *The Poor Law Magazine for Scotland*, 3 (1860–61), 148.
61 North Lanarkshire Archives CO1/37/56, 1870–75, no. 2, Applications for Relief, Dalziel, CO1 37 56, Application for relief from Thomas Campbell. For other cases see ibid, no. 79, John Bryson and no. 80, David Brennan.
62 See, for example, North Lanarkshire Archives, CO1 50 22, New Monklands Parochial Board, Letter Book, 1847–49, case of William Craig, 15 May 1849 (miner given meat and clothes during time spent out of work due to illness); Mitchell Library, Glasgow City Archives, CO1 22 44, no. 124, Blantyre Parochial Board, Register of Poor, 1845–64, Alexander McCormick (paid for admission to Glasgow Infirmary).
63 Mitchell Library, Glasgow City Archives, CO1 27 90 no. 4, Parish of Carluke, Record of Application for Relief, James Watson, 18 July 1856.
64 Ibid., CO1 27 90 no. 141, Parish of Carluke, Record of Application for Relief, James Fyfe, 12 May 1858.
65 Ibid., CO1 27 90 no. 88, Parish of Carluke, Record of Application for Relief, George Moffat, 10 June 1857.
66 Sharon N. Barnartt, 'Disability as a Fluid State: Introduction' in Sharon N. Barnartt (ed.), *Disability as a Fluid State* (Bingley: Emerald, 2010), 1–22.

67 Mitchell Library, Glasgow City Archives, CO1 27 90, no. 194, Record of Application for Relief, Thomas Kerr, 10 December 1859; CO 27 90 no. 199, Record of Application for Relief, John McKendrick, 27 December 1857.
68 Ibid., CO 27 90 no. 329, Record of Application for Relief, Samuel Donaldson, 31 March 1862.
69 S. A. King, 'Constructing the Disabled Child in England, 1800–1860', *Family and Community History*, 18 (2015), 104–21.
70 John Benson, 'Poor Law Guardians, Coalminers, and Friendly Societies in Northern England, 1860–1894: Statutory Provision, Local Autonomy, and Individual Responsibility', *Northern History*, 44:2 (2007), 164.
71 Glamorgan Archives, P62/28, Llantrissant Parish Vestry minutes, Overseers' and Churchwarden's Accounts, 1813–25, 319.
72 Benson, 'Poor Law Guardians', 161, 164.
73 Ibid., 160–1; Benson, *British Coalminers*, 175.
74 McCord, 'Relief of Poverty', 95. Eric Hopkins, *Working Class Self-Help in Nineteenth-Century England: Responses to Industrialisation* (London: UCL Press, 1995).
75 Benson, 'Thrift', 412.
76 George L. Campbell, *Miners' Insurance Funds, their Origin and Extent* (London: Waterlow and Sons, 1880), 8; *Merthyr Telegraph*, 4 April 1857.
77 *Colliery Guardian and Journal of the Coal and Iron Trades*, 27 September 1878, 497.
78 *The Times*, 26 May 1880, 14.
79 Anon. [Samuel Smiles], 'Workmen's Benefit Societies', *Quarterly Review*, 116:232 (October 1864), 318.
80 PP 1873 [C.842], *Third Report of the Commissioners Appointed to Inquire into Friendly and Benefit Societies; Together with Minutes of Evidence, Appendix and Index*, 170, 175.
81 Benson, 'Thrift'.
82 Hopkins, *Working-Class Self-Help*, 9, 12.
83 James C. Riley, 'Disease Without Death: New Sources for a History of Sickness', *Journal of Interdisciplinary History* 17:3 (1987), 554.
84 Martin Gorsky, 'The Growth and Distribution of English Friendly Societies in the Early Nineteenth Century', *Economic History Review* 51:3 (1998), 493–4; Dot Jones, 'Did Friendly Societies Matter? A Study of Friendly Society Membership in Glamorgan, 1794–1910', *Welsh History Review*, 12:3 (1985), 324–49.
85 PP 1834 (44), *Report from His Majesty's Commissioners for Inquiring into the Administration and Practical Operation of the Poor Laws*, 887.
86 Hopkins, *Working-Class Self-Help*, 46; Benson, *British Coalminers*, 184.
87 Edmund Stonelake, *The Autobiography of Edmund Stonelake*, ed. A. Mór-O'Brien (Bridgend: D. Brown and Sons, 1981), 55, 57, 58.
88 Hopkins, *Working-Class Self-Help*, 3.
89 For example, National Archives of Scotland FS4/124, St Aloysius Friendly

Society and Funeral Society, Chapelhall, Lanarkshire, 1863; ibid., FS4/101, Patna Loyal Orange Permanent Friendly Society, No. 103 Ayr, 1872–83.

90 Gwent Archives, D32.149, Rules of a Friendly Society of Tradesmen and others, called The Star Friendly Society, held at the Three Cranes, in the town of Pontypool, Monmouthshire (1831), rule 4. For later examples from other coalfields, see PP 1874 [C.961], [C.961-I], *Fourth Report of the Commissioners Appointed to Inquire into Friendly and Benefit Building Societies. Part I. Report of the Commissioners, with Appendix*, 181; Leslie A. Falk, 'Coal Miners' Prepaid Medical Care in the United States – and Some British Relationships, 1792–1964', *Medical Care*, vol. 4 (January–March 1966), 38.

91 TNA, FS2/3 Friendly Societies Indexes to Rules and Amendments, Series 1, no. 250, Beamish Colliery Friendly Society, 29 September 1835 to 5 September 1843; FS2/6 no. 57 United Colliers and Miners' Society, established 29 December 1831; National Archives of Scotland, FS1/16/20, Regulations of the Carfin Colliery Friendly Society, Instituted 12 October 1839.

92 TNA FS1/120 (Durham) no. 235, Rickleton and Harraton Outside Collieries' Relief Fund, certified 23 September 1833; Benson, *British Coalminers*, 179–81; Hopkins, *Working-Class Self-Help*, 44; P. H. J. H. Gosden, *The Friendly Societies in England 1815–1875* (Manchester: Manchester University Press, 1961), 86.

93 Steve King, '"It is Impossible for our Vestry to Judge his Case into Perfection from Here:" Managing the Distance Dimensions of Poor Relief, 1800–40', *Rural History*, 16:2 (2005), 161–89.

94 Simon Cordery, *British Friendly Societies, 1750–1914* (Basingstoke: Palgrave Macmillan, 2003), 26.

95 PP 1847–48 (648), *Provident Associations Fraud Prevention Bill. Report from the Select Committee of the House of Lords ... Together with the Minutes of Evidence and Appendix*, 62

96 Frederick Morton Eden, *Observations on Friendly Societies, for the Maintenance of the Industrious Classes, during Sickness, Infirmity, Old Age, and Other Exigencies* (London: J. White and J. White, 1801), 19–21.

97 Gwent Archives, Q.FSR.24–11, Rules and Orders of the United Colliers and Miners Society, meeting at the Clarence Inn[,] Trosnant, near Pontypool, established 4 September 1828, article 15; Swansea University, South Wales Coalfield Collection [hereafter SWCC], MNA/NUM/1/34/161, Glais Friendly Society, Rules and Articles, 1809, rule 42.

98 Hopkins, *Working-Class Self-Help*, 28–9.

99 James C. Riley, *Sick, Not Dead: The Health of British Workingmen during the Mortality Decline* (Baltimore: Johns Hopkins University Press, 1997).

100 *Colliery Guardian*, 12 September 1863, 204. For further evidence of the presence of the affiliated orders in mining settlements, see TNA, FS 2: Friendly Societies Indexes to Rules and Amendments, Series I, particularly FS2/3, FS2/8 and FS 2/13.

101 PP 1852–53 (955), *Friendly Societies. Return to an Order of the Honourable*

House of Commons, Dated 23 August 1852; for Copy of a Report and Tables ... on the Subject of Sickness Among the Members of Friendly Societies, as Shown by the Quinquennial Returns, to the 31st Day of December 1850, xxvi. For examples of age restrictions, see TNA FS1/120 (Durham) no. 231, Hetton Colliery Agents and Workmen's Friendly Society, certified 10 April 1832, rule 3; Swansea University SWCC, MNA/NUM/1/34/161, Glais Friendly Society, Rules and Articles, 1809, rule 23; Gwent Archives, Articles of the New Union Society, held at the Cross Keys [Inn] Pontypool, Monmouthshire, 1815, article 3. Benson, 'Coalowners, Coalminers and Compulsion', 55.
102 PP 1849 (458), *Report from the Select Committee on the Friendly Societies Bill*, 69; see also PP 1861 (464), *Friendly Societies. Report of the Registrar of Friendly Societies in England*, Appendix, 56.
103 Glamorgan Archives, DODD/6/2 Independent Order of Oddfellows: Bute Lodge (Llantrissant) Minute Book 1869–77, Lodge Meetings 11 October 1873, 8 November 1873.
104 PP 1872 [C.514], [C514-I], [C.514-II], *Second Report of the Commissioners Appointed to Inquire into Friendly and Benefit Building Societies, Part 1*, 59.
105 F. G. P. Neison, 'The Influence of Occupation on Health', *The Foresters' Miscellany*, April 1874, 81.
106 David Tonks, 'A Kind of Life Insurance: the Coal-Miners of North-East England 1860–1920', *Family and Community History*, 2:1 (1999), 48–9.
107 Cordery, *British Friendly Societies*, 26; Hopkins, *Working-Class Self-Help*, 18–20.
108 For example, National Archives of Scotland, FS1/16/29 Carfin Colliery Friendly Society, 1839; Gwent Archives, Q.FSR.14–13, Rules and Orders to be Observed by a Friendly Society of Tradesmen and Others called the Sons of the Globe in Pontypool, Commencing 21 March 1817, article xiii.
109 National Archives of Scotland, FS1/2/41 Kilmarnock Coal Cutters Society, 1834, article 2.
110 Gwent Archives, Q.FSR.14–13, Rules and Orders to be Observed by a Friendly Society of Tradesmen and Others called the Sons of the Globe in Pontypool, Commencing 21 March 1817, articles vii, xxvi, xxviii.
111 Gwent Archives, Q.FSR.16–4 Rules and Orders of a Friendly Society of Tradesmen, Coal Miners and Others, article 6.
112 PP 1842 (381), 513.
113 Deborah Stone, *The Disabled State* (Philadelphia, PA: Temple University Press, 1984), 99–103.
114 National Archives of Scotland FS4/1102, Rules of the Tranent Miners New Friendly Society, Tranent, Haddingtonshire, 20 October 1873, clause xxxv.
115 Swansea University SWCC, MNA/NUM/1/34/159, Glais Friendly Society Correspondence re Sick Benefit Claims, 1814–77, Folder 3, 1840–46, Letter 2, 22 June 1844.
116 Ibid., Folder 13, 1857, letter 12, n.d.
117 Martin Gorsky, 'Friendly Society Health Insurance in Nineteenth-Century

England' in Martin Gorsky and Sally Sheard (eds), *Financing Medicine: the British Experience Since 1750* (London: Routledge, 2006), 154–7; TNA FS1/120 (Durham) Friendly Societies Index to Rules and Amendments, no. 240, Hebburn Colliery Relief Fund, certified 15 February 1837, rule 17.
118 Gosden, *The Friendly Societies*, 138–40; Riley, *Sick, Not Dead*, 10 and ch. 4.
119 Gwent Archives, Q.FSR.14–1, Rules of the Loyal General Picton Friendly Society, Founded 12 August 1817, rule 13; ibid, Q.FSR.16–8, Monydduslywyn [*sic*] Monmouthshire, Rules and Orders to be Observed by a Society of Tradesmen ... Called the Faithful Britons, Commencing October 3 1818 rule 30.
120 Riley, *Sick Not Dead*, 50; 'The Sick Chamber', *The Foresters' Monthly Miscellany*, vii, April 1848, 161–2.
121 National Archives of Scotland, FS4/1228 Rules of the Wishaw Iron Works, Miners' Annual Friendly Society, 1874, article xvi.
122 Riley, *Sick, Not Dead*, 18, 37, 101.
123 Gwent Archives, Q.FSR.13–14 Rules of a Friendly Society Held at the House of Edward Duffield in the Parish of Risca (Founded 11 May 1805), 14 July 1817, rules 12, 17.
124 Gwent Archives, D2174.262, The 'Sons of the Globe' ... Treasurer's Account Book, 1818–68, Disbursements 1820/1.
125 Jules Ginswick (ed.), *Labour and the Poor in England and Wales 1849–1851: the Letters to the Morning Chronicle from the Correspondents in the Manufacturing and Mining Districts, the Towns of Liverpool and Birmingham, and the Rural Districts*, vol. 3: *South Wales–North Wales* (London: Frank Cass, 1981), 85.
126 National Archives of Scotland, FS1/16/29, Regulations of the Carfin Colliery Friendly Society, Instituted 12 October 1839.
127 Ginswick (ed.), *Labour and the Poor ... Vol. 3*, 56.
128 PP 1873 [C.842], 131.
129 For example, Rickleton and Harraton Outside Collieries' Relief Fund, Rule 5.
130 PP 1842 (008), 85, 102; Riley, *Sick, Not Dead*, 17.
131 See, for example, the rules of the Carron Friendly Society: PP 1874 [C.961], [C.961-I], *Fourth Report of the Commissioners Appointed to Inquire into Friendly and Benefit Building Societies. Part I*, 404–5.
132 Glamorgan Archives, DODD/6/1 Independent Order of Oddfellows: Bute Lodge (Llantrisant) minute books, lodge meeting 24 January 1857.
133 Benson, 'English Coal-Miners' Trade-Union Accident Funds'.
134 'Assuring Colliers' Lives', *Paisley Herald and Renfrewshire Advertiser*, 10 July 1869.
135 *The Miner and Workman's Advocate*, no. 16, 20 June 1863, 4.
136 Campbell, *Miners' Insurance Funds*, 6, 8. Benson, 'Coalminers, Coalowners and Collaboration', 183.
137 'Miners' Permanent Relief Fund', *The Newcastle Courant*, 8 May 1863.
138 'Northumberland and Durham Miners' Permanent Relief Fund', *The Newcastle Courant*, 13 May 1864.
139 'The Miners' Permanent Relief Fund', *The Newcastle Courant*, 19 June 1868.

140 PP 1873 [C.842], 126.
141 Ibid.
142 Ibid., 131.
143 'Proposed Miners' Permanent Relief Fund', *The Glasgow Herald*, 13 March 1878.
144 Campbell, *Miners' Insurance Fund*, 12.
145 Tyne and Wear Archives, CH.MPR/14/1 (MF2196), Miners' Permanent Relief Disablement Fund (1863–91).
146 'Miners' Permanent Relief Fund', *The Newcastle Courant*, 9 July 1880.
147 Tyne and Wear Archives, CH/MPR/1/1, Northumberland and Durham Miners' Permanent Relief Fund, committee minutes, 6 June 1863, 31 December 1864.
148 Tyne and Wear Archives, CH/MPR/5/6, Northumberland and Durham Miners' Permanent Relief Fund, Summary Report of the Sixteenth Annual Meeting, 13 July 1878.
149 'Northumberland and Durham Miners' Permanent Relief Fund', *The Newcastle Courant*, 13 May 1864.
150 'Miners' Permanent Relief Fund', *The Newcastle Courant*, 9 July 1880.
151 'The Cost of Colliery Accidents', *The Times*, 30 April 1879.
152 Benson, 'Coalminers, Coalowners and Cooperation', 187–91.
153 *The British Miner and General Newsman*, no. 2, 20 September 1862, 3; *Miner and Workman's Advocate*, no. 15, 13 June 1863, 5.
154 'Miners' Permanent Relief Fund', *The Newcastle Courant*, 16 May 1862.
155 *British Miner and General Newsman*, no. 8, 1 November 1862; ibid., no. 9, 8 November 1862, 3.
156 PP 1873 [C.842], 131.
157 Benson, 'Coalminers, Coalowners and Cooperation', 183.
158 Cordery, *British Friendly Societies*, 102.
159 Handel Cossham, 'Eminent Miners', lecture delivered to the Bristol Mining School, 3 March 1857 in *Lectures Delivered at the Bristol Mining School, 1857* (Bristol: Bristol Mining School, 1857), 164–5.
160 Benson, 'Thrift'.
161 On the broader context, see Stone, *Disabled State*, 32; David M. Turner, '"Fraudulent" disability in historical perspective', *History and Policy* (February 2012), http://www.historyandpolicy.org/papers/policy-paper-130.html, accessed 9 October 2016.
162 On the 'right' to relief, see Lorie Charlesworth, *Welfare's Forgotten Past: a Socio-Legal History of the Poor Law* (Abington: Routledge, 2010).
163 Hopkins, *Working-Class Self-Help*, chs 4–6.
164 Campbell, *Miners' Insurance Funds*, 11.
165 Borsay, *Disability and Social Policy*, 22, 27, 32; Hutchison, *A History of Disability*, 275–6.

4

DISABILITY, FAMILY AND COMMUNITY

The risks of coalmining affected not just the working lives of British miners during the nineteenth century, but also their lives beyond the pit. Many contemporary commentators sought to interpret the experiences of miners and their communities through the prism of their susceptibility to danger in the workplace. For example, in his comparative statistical study of Britain's 'dangerous classes', *Tactics for the Times* (1849), Jelinger C. Symons calculated that rates of criminality were lower in the mining districts of Northumberland, Cumberland, Durham and Cornwall than in other industrial areas or London. Compared to a national average of 28 out of every 10,000 persons committing property offences, the rate in these districts was merely 7 out of 10,000. This fact was explained by a relative lack of large towns in mining districts; the 'primitive and simple habits' of mineworkers and their families; and, above all, the constant 'peril to life' in underground labour, which served as a 'quickener to the moral sense'. To 'no class of men', wrote Symons, is the 'barrier between life and death slighter than among pitmen', and consequently there was an 'awe, partly religious, and greatly superstitious' that 'obtains amongst the people and check[s] vice'. In mining areas, he claimed, children were 'less lawless, and more subordinate to parental control', and women too were less liable to the demoralisation found in the cotton manufacturing districts of north-west England.[1]

Symons' association between 'uncertainty of human life caused by the frequency and terrible nature of accidents in mines' and low levels of theft in mining districts seemed 'fanciful' to other observers, who noted that criminality was high among other occupations who faced risk of accident and injury in their work, such as sailors.[2] Furthermore, in areas dominated by the iron trade, where ironworks and collieries existed side by side, rates of crime were notably higher.[3] However, Symons' attempt to provide links between miners' work, their exposure to risk and other aspects of their lives, was not uncommon. The historian K. S. Inglis explained that the success of Methodism in the mining communities of nineteenth-century England and Wales was due to the belief that 'miners lived closer to death' than other workers, and for the converted, experiences of loss and survival provided vivid evidence of the fragility of life and the importance of faith.[4] The rhetoric of pitmen turned preachers such as Richard Weaver was intimately shaped by their exposure to the risk of accident when working underground, describing the torments of the damned with 'an imagery gathered from the dense darkness of the coal-pit, the flames of the fire-damp, and the suffocating vapour of choke-damp'.[5] Yet if accidents reinforced the moral character of some, in others it produced a recklessness that found expression in a love for gambling and contributed to the stereotype of the miner as thriftless and lacking forethought.[6] Miners' marriages, alleged the author of 'The Collier at Home', an article published in *Household Words* in 1857, were 'founded on a rough sort of calf-love', in which there was little consideration for 'community of interest or feeling'.[7] To this author, proximity to danger produced an emotional detachment in miners' work and home lives, where the spectacle of a 'companion, burnt, or maimed or killed' merely lifted the 'deadly monotony' of labour at the coalface, and where grief for the dead may have been 'bitter in the first few days' but was speedily forgotten.[8]

When social commentators pronounced on the impact of risk on miners' social relations and emotional lives, they had in mind primarily miners' exposure to *fatal* accidents. However, as we have seen, non-fatal accidents were far more common. The living legacy of mining's dangers was visible for all to see in coal communities in the maimed bodies of survivors. While walking down a busy street in mid-century Merthyr Tydfil, the *Morning Chronicle*'s correspondent observed that there were 'more men with wooden legs than are to be found in any town in the kingdom having four times its population' – a consequence of the great 'number of accidents in the works below and above ground resulting in amputation'.[9] How were people with impairments viewed in coalfield communities; how did they regard themselves; and what social roles did they play? This chapter examines how social relations in mining areas were shaped by disability and asks how the lives of men, women and children were affected by impairments or chronic illness – whether their own, or those of family members. Despite significant research on evolving patterns of home life, leisure and religion in the coalfields, there has been little attempt to examine how social and familial relations of miners, and their emotional or spiritual attachments, were affected by illness or impairment.[10] This chapter contributes to our evolving understanding of coalfield life by situating the disabled miner within three distinctive, but overlapping settings:

in the community; at home; and in the religious activities of mining areas. It explores the ways in which impairment became visible in these settings and how the norms and values associated with these arenas both delineated the experiences of disabled mineworkers and were challenged, modified and redrawn after disablement.

The miner in his community

In spite of the image of the isolated, close-knit pit village, nineteenth-century coal communities were not homogeneous. Mining settlements varied considerably in their layout, quality of facilities and amenities, degree of isolation and in the extent of their dependence on coalmining alone.[11] The expansion of the coal industry led to considerable migration into mining areas, particularly from the surrounding countryside. The population of mining communities ebbed and flowed according to the demands for coal. Colliers were diverse in terms of ethnicity and religious affiliation. In 1891, for instance, the Durham mining village of Seaton Delavel was home to people born in 118 different towns and villages.[12] Migration, as we shall see in the following chapter, increased during times of strikes. Some mining areas, particularly in south Wales and Scotland, had significant Irish populations.[13] Over the course of the nineteenth century, as the industry expanded and communities became more established, so religion, leisure activities and educational pursuits served to bind together coal communities in stronger cohesion, although the unity of social and cultural values should not be overstated.[14]

While the structure of coal communities was diverse, contemporary images of the miner in his community tended to divide between two stereotypes. On the one hand, the miner had long been presented in popular culture as a harddrinking, raucous and irreligious character, spending his wages in merriment. This was a view cemented in eighteenth-century ballads, such as Newcastle poet Edward Chicken's *The Collier's Wedding*, originally dating from 1729, which celebrated a class of people who 'liv'd drunken, honest, working lives'.[15] The image of the carefree, hard-drinking miner, survived into the nineteenth century, but was beginning to be challenged by alternative views of the miner in his home and community.[16] Remarking on the moral state of English miners in his introduction to *The Pitman's Pay* (1843), Thomas Wilson remarked that 'the pitman's character has undergone considerable amelioration', since Chicken's time, a result of Sunday schools that increased children's literacy, the spread of useful knowledge in cheap publications and the introduction of savings banks which had produced 'care and economy among this invaluable class of men'.[17] An article about the Durham and Northumberland miners published in 1885 similarly noted that it was a 'mistake to suppose' that the typical miner lived life to the full 'while his wife and children starve'. By now, the pitman was 'frequently a teetotaller, and has no more favourite place of occupation for his leisure hours than the reading-room or the mechanic's institute'.[18] The improvements brought about by education and religious instruction were underpinned by an ideal of physical and moral fitness.[19] 'Veritas', a correspondent to the *British Miner* in 1862, celebrated the end of violent and cruel sports among miners of Northumberland and their replacement with 'cricket, bowling, running, etc, all of which help to develop their muscular strength'. 'Physical exercise', he continued, 'has made Englishmen what they are'.[20] The expansion of sport and leisure activities, chapels and institutions of learning, whether schools or miners' institutes that nurtured auto-didacticism, as coalfield communities became established all contributed to a reform of popular culture in which 'the humanising influences of religion, science and literature' challenged the image of the coalminer as a drunken hard man.[21]

Nevertheless, the raucous image of the miner in his community was not entirely displaced; nor was it confined to able-bodied men. Mid-Victorian Merthyr Tydfil's conspicuous population of amputee coalminers sometimes made themselves visible through acts of drunken violence, which made good copy in local newspapers. The *Merthyr Telegraph* published a number of reports taken from the town's police court in the 1860s and 1870s, detailing the aggressive behaviour of men with wooden legs, many of them former coalminers. In the summer of 1862, for example, an amputee collier named Henry Williams was charged with assaulting two police constables after they had tried to stop him beating his mother. Williams apparently threatened the officers, boasting that he had 'beat five policemen at Aberdare', and kicked at them with his wooden leg, tearing the coat and trousers of one of them. He was fined, ordered to pay fourteen shillings compensation for damage to the policeman's uniform, and sentenced to a month's imprisonment with hard labour.[22] A year later, John Evans, a collier with a wooden leg, was charged with 'being drunk and fighting in the public streets', while in 1873 Morgan Price, alias 'Mockyn Croes-pen', another wooden-legged collier, was accused of 'pugilism' after staging a brawl on the highway near Cefn-coed-y-cymmer on the north western edge of the town.[23] Stories such as these simultaneously contributed to the rowdy image of this Welsh industrial town, which was notorious for drawing in disorderly people from the surrounding area and further afield, while also attesting to the attempts of the authorities to clamp down on unacceptable behaviour.[24] However, stories of violence involving disabled miners were found in other coalfields as well. In 1875, for example, the *Glasgow Herald* reported the trial of a 'blind brawler', collier James Marshall who was 'totally blind', accused of causing a disturbance in a public house at Rawyards, Airdrie, after being 'repeatedly convicted of a similar offence'.[25]

These stories of drunk and violent impaired colliers can be interpreted in various ways. On the one hand, they can be read as stories of displacement, in which men with significant impairments were reduced to the role of dangerous outsider, threatening the stability of their communities. They appeared at a time when the temperance movement was gaining ground in industrial south Wales and other parts of the country, and when working-class male respectability was becoming firmly associated with the ideal of the breadwinner who provided for his family rather than squandering money in the pub. In this context, the violent and disorderly amputees depicted in these reports represented the antithesis of ideals of sober manliness.[26] But at the same time, these stories show ways in which men whose livelihoods and status were threatened by impairment might fall back on the image of the tough, hard-drinking miner as a means of rejecting any associations between physical impairment and vulnerability or weakness. These were 'disabled' men determined to demonstrate their physical strength, whatever their impairment. They appear as not just getting into trouble, but positively *inviting* it, seeking opportunities to test their strength against able-bodied opponents. As Shani D'Cruze has noted in her study of crime in Victorian England, '[m]en's reputation as fighters could form an important component of their masculine self-respect.'[27] Modern research shows that men with impairments acquired after birth sometimes attempt to defend their masculine identities from the potentially feminising threat posed by disability by 'proving' their manliness through acts of aggression or physicality.[28] With these insights in mind, the violence of men like Henry Williams and Morgan Price can be seen as a brutal assertion of their manhood.

By including details of brawling disabled miners' impairments, journalists gave their stories about these men a 'freakish' quality, inviting readers to reflect on the surprising capabilities of amputees and other visibly injured people, or view them as figures of fun. In common with other Victorian newspaper accounts of violent crimes, many reporters emphasised the physical impairments and deformities of defendants to portray the accused as a 'comic grotesque'.[29] A story printed in the *Merthyr Telegraph* about another coalfield amputee, David Thomas, who tussled with a police officer, appeared under the headline 'A Dangerous Leg', and described how Thomas had 'used his wooden leg with *extraordinary proficiency*', giving the constable a 'punch' on the shin, in such a way to 'remind him that limbs of flesh were not always the most formidable weapons in a scuffle with a drunken man'. In the account of his trial, Thomas was described mock-heroically as 'the hero of the wooden leg'. The newspaper reported laughter when the court told the defendant to 'learn to keep that wooden leg of yours in reserve for its proper purpose', and further laughter when he was warned not to 'kick out with such a dangerous limb as that, or you may find yourself kicking some one's brains out some of these fine days'. Although found guilty of a serious offence, Thomas's treatment by the court and in the press report repeatedly used his impairment to reduce the case to farce.[30]

While some impaired colliers were seen as dangerous and disruptive presences in their communities, others were adopted more affectionately as local 'characters', whose physical impairments were part of a set of eccentricities that added to their distinctiveness and appeal. As James Gregory has argued, the 'eccentric biography' was a popular genre for writing about physical difference in eighteenth- and nineteenth-century Britain and these narratives often imbued those with physical imperfections with compensatory mental qualities of character and personality.[31] Such was the case in the *Merthyr Telegraph*'s affectionate portrait of William David Richards, 'better known far and wide as Billy Davy Richards' the 'Bard of Pen-Heol-Gerrig', published posthumously on 29 May 1858. The narrative placed Richards, who wore a wooden leg for most of his life following a mining accident, at the centre of his community, a schoolmaster and bard, a man who acted as secretary to 'several clubs', who would spend his evenings smoking a pipe, discussing 'politics or gossip, bardism or antiquity, with a select knot of friends'. Richards was regarded as the 'oracle of the stony hamlet wherein he lived' and was consulted on all matters, taking up the cause of his neighbours in letters and petitions, particularly to the Crawshay family who owned the local collieries and iron works.

'None who knew him', wrote the author of his life story, 'would deny him the possession of virtues and of traits worthy of esteem, but he was frail.' Dying exhausted and emaciated by poverty, 'poor Billy' provided an example of determination and bearing with misfortune. Using a horticultural metaphor that at once captured his deformed body and his steadfast endurance, the author described him as being like 'one of the gnarled and wiry, the storm bent and dwarfed oaks upon our hill sides' which 'endures for an age', contrasted with the 'well grown and slender poplar' which was 'too often preferred' on the basis of its appearance, yet was merely a 'thing of the day'. In this account, Richards was presented not so much as someone who achieved much in his life *in spite* of his impairment, but rather as one whose physical difference had shaped his mental outlook in a positive way, giving him the qualities that were admired by those who knew him, but which went unappreciated by those who judged on appearances alone. Like the 'dwarfed oak', there were 'inherent powers within that rugged mass that would have displayed themselves in the stately form and ample proportions of the monarch of the woods', if only they were planted 'in better soil, a different location, and under different circumstances'.³² Richards' misfortune lay in being born into a community where making a living for men was associated with physical prowess, which meant that the many compensatory qualities that made up for his broken body did not lead to the success he deserved.

The image of nineteenth-century mining towns and villages as cohesive settlements found its principal expression in the uniting of the community in worry and grief in the wake of accidents. An account of the aftermath of an explosion of fire-damp at Jarrow Colliery in 1826, published in the *Methodist Magazine*, was typical of many in its depiction of the 'whole neighbouring population' being 'drawn together' in the minutes and hours that followed the disaster, as rescuers searched for survivors and the dead were brought to the surface:

> The cries and lamentations of the distressed relations, the suggestions and opposing shouts of the persons busying themselves in the work, the blazing of the surrounding fires, and the uncertainty of the extent of the destruction, conspired to fill every breast with terror and distress.³³

Bringing home the dead, and funerals of the victims of fatal accidents, were depicted as public events in which the grief of widows and children was shared by all.³⁴ Nevertheless, if fatal accidents were occasions for unity that bonded communities through sharing the burden of loss, the willingness of coalmining communities to embrace those left disabled by accidents, or people with disabilities in general, depended on a variety of factors. The mutualism spurring the worker-led welfare schemes examined in the previous chapter promoted the idea that local communities ought to support disabled workers. Yet this was not always the case in practice and attitudes towards disability in coalfield society could be indifferent, or openly hostile. Sympathies for the maimed were usually strongest immediately after an accident, but often waned over time. Jim Bullock's memories of growing up in a Yorkshire mining community around the turn of the twentieth century included seeing a 'once-strong miner', now paralysed with a broken back, lying on a water bed in the street 'where his relatives had wheeled him to get some sunshine and chat to the miners'. In the early days after his accident 'all his friends and relatives were helpful and sympathetic', but as time wore on and his condition did not improve, he became less interesting to others and increasingly aware of himself as a 'burden' and so he 'began to grumble and curse about the cruel blow struck him in the prime of life' – 'His complete manhood had been taken away from him.'³⁵

Although disability was normalised in mining communities of the nineteenth century, people with disabilities might on occasion be subject to cruelty and laughter. On 6 May 1865, the *Aberdare Times* reported the pastime of 'Hunting a Packman', a game in which a female 'cripple' was chased by women and 'urchins' through the streets – she giving them 'knocks and tumbles' with her crutch and the chasers trying to trip her up. 'All seemed to enjoy it', reported the paper, although the views of the 'packman' herself were not recorded.[36] A 'packman' was a pedlar or itinerant trader, and the cruel treatment she received may have derived as much from the fact that she was a stranger in the neighbourhood, and possibly taken for a beggar, as from her physical difference. Social attitudes towards the disabled poor in mining communities were laced with concerns about imposture. The high number of 'cripples' produced in the expanding coal industry may have drawn beggars to some mining communities in the hope of getting assistance, heightening concerns about impairments being faked in order to gain sympathy.[37] While these concerns were universal during the nineteenth century and were centuries old, hostility to imposter beggars may have been felt particularly keenly in mining areas due to the high incidence of genuine impairment.[38] Accusations of fraudulent presentations of disability in mining and other industrial areas were often levelled at 'outsiders', reflecting tensions caused by migration in search of work. Irish migrants who came to coal and iron districts in Scotland and south Wales to supply shortages of labour during strikes were accused of using 'various and numerous devices to obtain assistance from the parish for their wives and families, whilst they are themselves at work'.[39] Such hostility helps explain why many Irish friendly societies were established in Britain's coalfields during the nineteenth century and suggests the role of ethnicity in shaping experiences of disability.[40] For outsiders seeking belonging and acceptance, mining settlements or neighbourhoods were often unwelcoming places. In such circumstances, sympathy and kindness could be hard to find, even in ill health.

'Strangers' presenting impairments who entered mining communities requesting alms were frequently met with suspicion, sometimes rightly. The Durham miner John Wilson described how the elder sister in the Todd family, with whom he lodged as a young worker at Ludworth Colliery around the middle of the century, had chased away a beggar feigning to be 'dumb' when he came to her house, exposing his imposture by beating him with a brush until he 'cried out for mercy'.[41] At Risca in Monmouthshire in 1862, David Miles, 'a tall powerful looking fellow', was arrested for begging, claiming that his arm had been 'smashed at a coal pit in Aberdare'. Upon examination by a doctor, his arm was declared 'perfectly healthy' and Miles admitted to feigning impairment explaining that 'his reason for doing so was that he could not support his family otherwise than by begging' and felt 'ashamed to beg' being 'an able-bodied man without an excuse'. He was sentenced to six weeks' hard labour, the judge declaring that 'these acts of gross deception must be punished, as they are calculated to deter persons from relieving the really necessitous'.[42]

The judge in this case made a good point: community fears about fraudulent disability influenced the experiences of people with real impairments, often in negative ways. Hoping for sympathetic treatment, or at least respect, some people with genuine health problems encountered derision or outright animosity in their daily lives. In his autobiography, Edward Rymer recalled the time he visited north Wales to promote trade unionism among coalminers there in the late 1860s. During one of his speeches, some members of the audience, Rymer thought, 'grinned and laughed at my lameness'.[43] Rymer has been characterised as a divisive, controversial and difficult figure, sensitive and easily offended.[44] Coupled with the fact that he led a peripatetic life, this may have compounded his 'outsider' status in the places he lived and visited, exacerbating any prejudice and ridicule he experienced because of his impairments. He certainly claimed to have been hurt or angered on other occasions by comments allegedly made about his physical appearance or abilities.[45] If Rymer's personality and lifestyle contributed to the frequently harsh attitudes of others towards him, however, his belief that he faced prejudice because of his impairments was echoed in other eyewitness accounts of coalfield life. Born in 1835, James Dunn worked as a mineworker in the English Midlands as a child. During that time his health was badly affected by a tumour that eventually forced him to leave mining. Reflecting on his childhood many years later, Dunn referred to his tumour and highlighted it when he wrote that the 'poverty and affliction' he suffered 'excited pity in some hearts, but the opposite feelings in others'.[46] Like poverty, illness and injury provoked mixed responses and feelings within coal society. Despite romanticised portrayals of steadfast miners rallying together to support workmates in need, disabled mineworkers could not *count* on the acceptance and kindness of their communities.[47]

Home life

In January 1873 the *Illustrated London News* (*ILN*) published an illustration of an 'interior of a collier's cottage' (Figure 3). The picture depicts a Welsh collier and his family inside their home. The collier, a serious looking man who holds a pick, is at the centre of the picture, standing close to a fireside in what is presumably the kitchen. Beside him are two women, one on each side. One does the laundry while the other woman sits with an infant on her lap, another child close by. A dog sits at her feet looking devotedly towards the miner.[48]

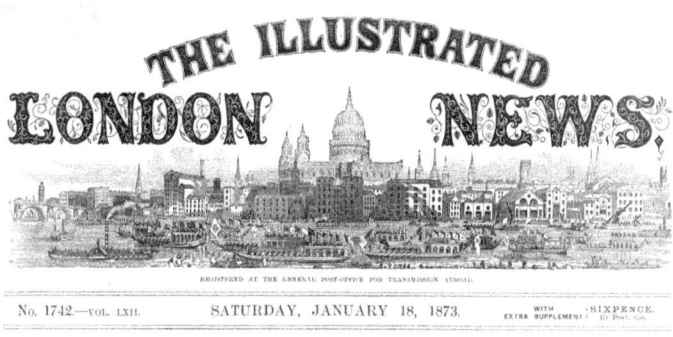

Figure 3 'The Strike in South Wales: Interior of a Collier's Cottage', *Illustrated London News*, 18 January 1873. Copyright Illustrated London News Ltd/Mary Evans.

Obviously an idealised scene, the picture does, however, reveal something of the reality of nineteenth-century miners' domestic lives. As its composition implies, mining was at the heart of mineworkers' households, influencing what they did and how they developed.

Mining households were essentially sites in which mining labour was reproduced, nurtured and maintained. Highly gendered, mining households' activities were geared towards meeting the fundamental needs of mineworkers so they could carry on the arduous mining labour on which their families relied financially. As suggested by the *ILN*'s picture, women kept the home and performed the vital domestic chores necessary to ensure miners were sufficiently nourished, clothed and cared for to continue working at their jobs. Women also bore, raised and tended the children that became the pit workers or miners' wives of the future. And so the cycle continued.[49]

The *ILN*'s picture also speaks to another feature of miners' lives, including those who today might be considered 'disabled': family was very important to their everyday experiences. As we have seen, despite the rise of specialist medical and welfare institutions in the nineteenth century, sick and injured miners at this time overwhelmingly lived in family settings, not institutions. The families in which they lived, however, could vary quite widely. Many types of working-class household existed during the Industrial Revolution, including those tightly focused on the nuclear family unit, others that were more fluid and contained non-nuclear kin and others unrelated to the nuclear unit, such as lodgers.[50] Mining families were no different.

A sense of the range of households in which disabled miners lived is found in Scottish Poor Law records and the census of 1871. When the census was taken, William Scott was in his mid-fifties and living in Carluke, Lanarkshire. Described as a miner unable to work, Scott was listed as living with his wife, Margaret (fifty-one), and his three children: Janet (fifteen), William (twelve) and Robert (ten).[51] A few weeks later, Poor Law officials confirmed his incapacity when they categorised him as 'wholly disabled' and granted him relief, noting he was 'suffering from Miner's Asthma'.[52] Fifteen-year-old Robert Hamilton was also living in Carluke in 1871 when census enumerators surveyed his household. Like Scott, he too was listed as a mineworker and Poor Law officers considered him 'wholly disabled' (being 'unfit for work from pain in [his] left side') shortly before the census. His household was much larger and more complex than the one enumerators found when they called on the Scott family, however. An orphan, Hamilton was recorded as living with nine other persons. Headed by James Brown, the household in which Hamilton lived also included Brown's wife and five children along with another teenaged orphan (named Jonas Hamilton and possibly Robert's brother) and William

Neilson, a twenty-four-year-old boarder.[53] Scott and Hamiltons' domestic circumstances indicate that both nuclear and non-nuclear households were found in coalfield communities and that these could vary quite considerably in size and composition. Like the Browns, many families took in lodgers or other people seemingly unrelated to them by blood or marriage, such as Neilson or the Hamilton boys. It was rare, though, to find non-nuclear households where nobody was related to their fellow co-residents.

Kinship ties underpinned the vast majority of mining households. At the time of his death in the Gethin Colliery explosion of December 1865, amputee Griffith Ellis lived together with his brother and sister in Abercanaid, south Wales. The brothers, in line with the gender ideology of the time, had been the household's main breadwinners, working for pay in the pits, while their sister worked at home 'keeping house for them'.[54] Co-operation and mutual dependence between men and women in this way was the norm in mining communities. Given that essential domestic tasks such as cooking, cleaning and child care were considered 'women's work', women were a constant and *vital* presence in mining families. Jane Humphries has observed that working-class widowers were more likely to remarry than widows. That they did so, she suggests, was partly because men realised they were so dependent on the domestic labour of women that it was almost impossible to maintain a functioning household without a woman to assist them.[55] Although domestic arrangements similar to the Ellis's were not unusual, most mining households had a husband and wife at their core. The families in which Robert Hamilton and William Scott lived in 1871, for instance, may have been quite different in terms of size and composition, but both centred on a miner and his wife.[56]

Although common-law unions or informal cohabitation were reported in colliery districts before the 1850s, and were believed to be common in some areas, marriage was the foundation upon which most mining households were built.[57] Men and women from mining families were noted for their tendency to 'marry among themselves' throughout the nineteenth century and this contributed further to the idea of miners as a 'distinct race of beings'.[58] Along with their endogamy, colliers also had a reputation for marrying young and having lots of children – a view supported by demographic evidence.[59] Miners were able to marry younger than other workers because of the relatively high wages they commanded, but their reasons for marriage were essentially the same as other labouring Britons'. While romance and sexual attraction may have played their part, the decision of working-class couples to marry or live together was frequently motivated by economic considerations.[60] For colliers, wives brought with them productive and reproductive capacities that were a significant boon, not only in the domestic sphere, but also in the workplace. Referring to the time when women worked at the Scottish mines he managed, John Wright claimed that men there 'married more from the advantage their [spouse's] physical strength might procure them, than any degree of affection'.[61] Before they were forbidden from doing so in 1842, miners' wives commonly worked underground in Scotland as coal bearers. By carrying the coal their husbands cut, wives enabled miners to spend more time at the coalface hewing, thereby increasing their earning potential. Even when not employed in mines themselves, wives still provided colliers with valuable mining labour by giving birth to their children. Miners' sons, particularly after the prohibition on underground female labour, were especially likely to follow their fathers into the mines.[62] As we saw in the previous chapter, the income child mineworkers earned was crucial for the economic survival of many mining families. Wives and children were materially advantageous to miners in other ways too. In coal districts where company housing was common, the allocation of homes was often determined by the marital status of mineworkers and the size of their families. In north-east England, for instance, married men with lots of children received preferential treatment in this regard and were far more likely to get company accommodation than single men or those with small families.[63]

As in the workplace, injury and ill health were ubiquitous in mining families. We have already observed the households in which 'disabled' mineworkers such as Griffith Ellis, William Scott and Robert Hamilton lived. Men and boys were not the only people in coalfield society with physical impairments, of course. As we saw in Chapter 2, prior to 1842 reformers calling for the prohibition of women and girls from working underground highlighted their susceptibility to ill health and disability. Furthermore, coal districts also included many people with congenital impairments or conditions who depended on the income of mineworkers for support, like the teenage 'dwarf ... in very bad health' mentioned in a letter detailing the families of dead mineworkers killed in the Gethin Colliery disaster of 1865.[64]

The prohibition on women's underground labour in 1842 served to consolidate the male breadwinner ideal. While some miners' wives supplemented household income through economic activities of their own, such as running a 'small huckster's shop' selling food and other small articles, in many coal communities after 1842 married women's work was more likely to be centred on the home than in other working-class areas.[65] The landlady duties women performed for lodgers, for example, undoubtedly contributed greatly to the financial well-being of mining families, but this work was essentially an extension of the housekeeping responsibilities expected of wives and mothers *within* their households, not a foray into the world of work beyond.[66]

With the gradual consolidation of the idea that a woman's rightful place was in the home and the ban on females working underground, the gendering of industrial disease and disabling injury as *male* also became more pronounced as the period wore on. Despite this, women continued to be hurt in the service of the coal industry. Their contribution to mining through the work they did at home supporting miners took a toll on their bodies that could be just as incapacitating as work in collieries. In mid-Victorian Merthyr Tydfil, for example, poor water supply there meant colliers' wives had to perform hard labour on a daily basis when they carried water long distances from the River Taff to fill their husbands' baths.[67] John Liefchild similarly observed the arduous nature of women's lives in northern coalfield settlements. 'The duties of a pitman's wife', he wrote, 'are very numerous.' Due to the fact that mineworkers in their families often worked different shifts, many women were busy 'preparing numerous meals ... at irregular and various hours of the day'. They also had to wash the incredibly dirty work clothes of their mining menfolk. All this activity, lamented Leifchild, echoing earlier critics of women's underground labour, left a woman 'with little time to attend to her duties as nurse' to her children.[68] Miners' wives were continually having to juggle the competing demands of childcare with looking after their fatigued husbands and mine-working sons. Usually they managed it, but often at a cost to their health. With the expansion of the coal industry, moreover, the physical burden women were under only increased.[69] Writing about the famous Welsh coal-producing communities of the Rhondda Valley in the late-nineteenth century, Dot Jones has argued that the 'unremitting toil of childbirth and domestic labour killed and debilitated Rhondda women as much as accident and conditions in the mining industry killed and maimed Rhondda men'.[70]

Impairment was undoubtedly a part of life for many people in British mining communities, but what impact, if any, did it have on the households in which they lived? The social, economic and emotional consequences of disablement to the family lives of miners are often elusive. Reconstructing them requires careful piecing together of fragments of information from newspapers, official reports and autobiographies. While it is always difficult to document historic experiences of disability completely, the available evidence suggests that although impairment might re-draw domestic relations in quite profound ways, the experiences of mining families varied considerably.

In the first place, there can be no doubt that disablement could be challenging for all families, not simply because of the potential loss of income, but also because of the emotional strain it placed on individuals and their relatives. Some disabling conditions, to which miners were susceptible, such as lung diseases, might be distressing and disruptive for caregivers as well as patients.

Lung diseases were typically accompanied by violent coughing, expectoration and other physical effects such as 'groaning'. Symptoms meant that patients developed their own routines that might be at odds with the established rhythms of the household. For example, Thomas Lewis, a fifty-two-year-old former miner from Aberdare in south Wales, suffered from asthma that 'compelled him to spend whole nights kneeling by the fire instead of going to bed'. One morning in March 1839, after some 'cross words', his wife Mary had struck him with a hammer, killing him. Mary Lewis's actions were attributed to insanity, believed to have been hereditary in her family and which had commenced in her after the birth of her last child some four years previously, but it is possible that the strain of caring for an incapacitated husband and several children could have exacerbated the situation.[71] In some cases, fear of becoming a burden was such a source of unhappiness for injured miners that they too felt the need to take similarly extreme measures. According to a newspaper account published in 1855, a thirty-year-old miner, Robert Perrie, employed by the Eglington Iron Works company in North Ayrshire, Scotland, who had become incapacitated from working thanks to a crushed ankle, shot himself through 'despondency ... apparently on account of his inability to work, and finding himself thereby a burden on his parents'.[72] Such cases were infrequent, but they reveal deeper histories of familial relations, duties and feelings indelibly affected by impairment.

For unmarried men like Perrie, depending on parents was an important – and in his case apparently humiliating – source of support during incapacity. But for married men and fathers, as we saw in the previous chapter, reliance on the earning power of other family members, especially children, was crucial after impairment. In her study of child labour in the Industrial Revolution, Jane Humphries has employed the concept of 'breadwinner frailty' to explain why working-class children were sent to work. Families' dependence on adult male earners, she argues, made them economically vulnerable to breadwinner unemployment or incapacity. Consequently, when men were sick, injured or unemployed, families had a strong incentive to encourage or force their children into work. This affected family dynamics in a number of ways. Fit and healthy boys whose labour might help their families in times of need were valued, whereas others – girls and more fragile boys – might find themselves marginalised.[73] So 'useless are daughters considered – not being allowed to go to the pit to earn money', remarked a piece on 'Births in Colliers' Families' published in a Scottish newspaper in 1855, that miners were reported to prefer that their female infants were stillborn.[74] In John Saunders' fictional tale *Israel Mort, Overman* (1876), young David Mort, too weak to follow his father into the pit, found himself erased from view within the home. His father, 'on finding him unfit for pit work, apparently lost all recollection of his very existence; scarcely seeming to see him when he crossed his path or when he sat opposite him at meal-times'. Yet potentially there were opportunities for the 'weak' child too. David Mort was able to get a better education than his peers who were absorbed into mine work and dreamed of becoming a schoolmaster, until finally family circumstances meant that even he was sent to work underground, his childhood ambitions crushed.[75]

Injury and illness, then, had the potential to disrupt established relations and dynamics in mining families in quite profound ways. As Julie Marie Strange has demonstrated, masculine authority within the family depended on the 'fulfilment of provider obligations' and when 'provision faltered, the power dynamics of domestic life changed'.[76] In his autobiography, former Northumberland miner Thomas Burt recalled how he became 'the responsible head of the house' following the 'breakdown of [his] father's health' and withdrawal from mining in the 1850s.[77] When sons stepped in to take their incapacitated fathers' place as the main breadwinners in their families they sometimes usurped the status of household head too. Such an occurrence represented a significant realignment of power relations within a family, at least symbolically.

Coal workers who found their position in their households under threat as a result of impairment had to adjust to their new circumstances. In extreme cases, this might lead to violence. Writing in the 1840s of his experiences as a doctor in Tranent, Scotland, S. Scott Alison recalled the time he 'attended a young married collier under disease produced by debauchery'. Not 'very able to work', the man in question, Alison reckoned, did not do so 'for a year or two'. During that time, the collier 'remained at home' and was supported financially by his wife who apparently had a job at a nearby colliery. Instead of treating his wife with gratitude and affection for all her efforts on his behalf, however, Alison reported that the man was known to have 'grossly assaulted' her on at least one occasion after she returned home from 'a day's hard toil'.[78] Whether or not this abusive miner was ever punished for his domestic violence is not known. Other impaired mineworkers were brought before judges for physically attacking their spouses. In February 1861, for example, the *Durham County Advertiser* reported that Edward Rymer had recently appeared in court 'charged with assaulting his wife' and was '[b]ound over to keep the peace for three months'.[79] Humphries has found that working-class men in the nineteenth century 'who were unsuccessful in the world of work were inclined to brutality at home'.[80] Like the hard-drinking, fighting miners with impairments referred to previously, then, men who felt their manliness and position within their families called into question by their reduced or complete inability to work may similarly have resorted to violence as a way of reasserting their masculinity and dominance at home.

Nevertheless, we should not assume that impairment always led to a loss of status that demanded such drastic and desperate action. Neither did it necessarily change the gendered organisation of mining households in any fundamental way. As we have seen, impairment did not automatically equate to inability to work and many miners continued working at collieries despite serious injury. Men who took lighter work or alternative roles in and around the collieries found a means not just for earning a living, but also for potentially regaining some of the provider role expected of fathers. Such work may have been considered menial or low status, but 'boys' work' might at least enable men to maintain some self-respect.[81] Fund-raising 'gatherings' held in some communities to buy tools or goods to enable men incapacitated from mining to take up alternative work furthermore demonstrates the importance attached to maintaining impaired men's dignity through helping them to provide for their families.[82] As attempts made under the Poor Law to force men who deserted their families to support them suggest, impaired husbands and fathers were expected to maintain the role of family provider as far as possible. In 1857, for example, the Wolstanton and Burslem Poor Law Union in Staffordshire advertised a reward for the apprehension of William Wagstaff, a collier 'blind with one eye' with burn marks on his hands and face, who had absconded from his family leaving it in financial trouble.[83]

Even when impaired miners did lose their ability to provide for their households, impairment never totally erased the gendered expectations that shaped men and women's roles within families – not even in the most exceptional circumstances. Although critics often accused women workers in early nineteenth-century Scottish collieries of neglecting their domestic responsibilities, evidence from the 1842 Children's Employment Commission report suggests otherwise.[84] For example, although Margaret Boxter (or Baxter) seemed to transgress gender roles by becoming one of the few women to work as a coal hewer (a job that epitomised 'masculine' skill and strength) after her husband could no longer work underground due to his trouble breathing, she was still expected to attend to the running of her home. According to her ten-year-old daughter, who also worked with her in the mines, Boxter went to the pit at four in morning, leaving 'at mid-day to do work at home, as father is bedridden'.[85] Although she subverted the gender ideology of the time by working as a hewer, then, Boxter was still unable (or unwilling) to overthrow it completely by relinquishing her duties as a housewife.[86] Disability might challenge the gendered identities of family members in various ways, but gender ideologies were so entrenched and widespread that men and women were rarely able to escape the roles expected of them entirely.

It should also be remembered that, as important as work was, it was not the only way through which Britons 'performed' their masculine or feminine identities.[87] For working-class men, a capacity for (paid) labour may have been a measure of manhood, but so too was marriage and fatherhood and the head of household status that tended to flow from this. Although some might have resorted to violence, then, miners unable to work because of illness or injury had other options available to them than simple aggression to demonstrate their manliness. 'Fathering a child', as William Howard has observed, was particularly valorised as an indicator of manhood in mining communities, as it signalled '[s]exual experience' and 'vigour'.[88] Studies of sex and disability show compellingly that, despite popular stereotypes to the contrary, disabled people have sexual desires and often lead very active sex lives.[89] This is as true for the past as it is the present and is tangibly borne out by evidence relating to mining families. Ill or injured mineworkers continued to have children even if they had trouble working. While impairment could restrict mineworkers' ability to earn a living and fulfil the breadwinner ideal, it rarely deprived them of the ability or inclination to have sex. Although Margaret Boxter's husband had to give up mining because of respiratory problems, his shortness of breath seems to have been no hindrance to him impregnating her afterwards. According to Mary Boxter, his daughter, she was 'born two years after father ceased to work in the mines'.[90] Other mineworkers similarly fathered children after injury or the onset of chronic illness.[91]

Once they were married, most couples in mining communities seem to have stayed together in the face of impairment and many carried on doing one of the main things that couples were supposed to do: having and raising children together. Impairment could certainly change the power dynamics of mining families in profound ways, but it did not radically transform the basic form or function of miners' households. Neither did it inevitably lead to a loss of headship. If an incapacity for work may have forced some miners to relinquish their status as 'head' of their families to a son or other relative, this was not universally the case. No such fate appears to have befallen miner William Scott of Carluke in the 1870s when he was struggling with the restrictive effects of respiratory disease. Although his entry on the 1871 census indicates he was unable to work, enumerators still described him as the head of his family.[92] Census data, of course, as other historians have pointed out, reveal little about the real nature of the relationships that underpin everyday family life.[93] While census officials may have regarded Scott as the head of his family, this does not mean he occupied the position in the fullest sense of the term's meaning. It is possible, indeed even probable, that Scott relied as heavily on his wife and children for support as they did on him. After all, the fact that he turned to the Poor Law suggests he struggled to fulfil the role of family breadwinner. It is also possible that he delegated some of the implied responsibilities of headship to others. Despite these caveats, however, Scott's classification as the 'head' of his household by officials still held great significance to his masculine identity and how others saw him. Given that nineteenth-century Britons regarded headship as a key sign of adult manhood, enumerators' recognition of Scott as family head represented a significant symbolic validation of his masculinity.[94]

Without doubt, experiences of disability within mining families varied. They were affected by the degree of a person's incapacity, by gender roles and expectations, by the existence of others who could provide economic or emotional support and by the opportunities for returning to work outside the home (as examined in Chapter 1). They were also affected by a person's stage in the life cycle.[95] Those injured before they married might have had different experiences from older men who had wives and children to support. Given that marriage represented, above all else, the pooling of a couple's productive (and reproductive) resources, it is possible that injury, illness or disfigurement may have affected a person's ability to attract a spouse. Soon after his engagement to Sarah Bradshaw in the early 1850s, mineworker Richard Weaver broke his hand in an accident. A serious injury, staff at Manchester Infirmary feared for his life, but Weaver survived.[96] His prospects for marriage, however, hung in the balance. During his convalescence, his wife-to-be visited him one evening and they discussed their engagement. According to Weaver's recorded recollection of the incident: 'she told me that her friends in the factory had been trying to persuade her to give me up, on the ground that I should probably be a cripple for life'. Luckily for Weaver, Sarah Bradshaw ignored the advice of her friends, telling them 'I will marry him, even though I have work to keep him.' The couple were married in January 1853 and a few years later had their first child together. While the possibility that Weaver would be 'disabled for life' did not deter Sarah Bradshaw from becoming his wife, it did force her to reconsider her relationship with him and recognise that it might entail extra 'work' for her.[97] Moreover, the comments of her factory friends suggest that not all women in industrial communities were as willing as Bradshaw to accept suitors with impairments. Working-class women were aware that marriage was fundamentally an economic union and that husbands with restricted physical capacities might struggle to fulfil the breadwinner duties expected of them, thereby endangering family prosperity.

Women in mining areas were not immune from harsh judgements about their bodies either. Like other working-class women, a miner's wife's health was '[c]rucial to her status within the household'.[98] When they chose a wife, men too looked for spouses with the physical attributes necessary to meet the gender expectations of the time. Wives were supposed to bear children and 'keep house'. Anything that was perceived to impinge on their abilities to fulfil these roles reduced their potential attractiveness to would-be husbands. Physical appearance was particularly important for some men in this regard. On being asked, '[w]hich on the whole make the best wives, pit women or others?' one Lancashire mineworker in the 1830s was reported as saying: 'I would as soon have one that has not been in the pit; many of them are crooked with going into the pit' – a judgement that reflected both their perceived lack of beauty and a potential inability to bear children.[99] Although functional assessments of bodily capacity could affect both men and women's marriage prospects, women were perhaps under greater pressure to conform to aesthetic standards as well.

While impairment could make finding a spouse more difficult, it was rarely an absolute barrier to marriage. Like Weaver, many other mineworkers also managed to marry *after* they were seriously injured or showed symptoms of chronic illness. In June 1847, William Smith, a 'collier by trade' in his early forties who had 'lost the use of his arms by an explosion in a coal-pit', died in Newbridge, Wales. Speaking at the inquest into her husband's death, Barbara Smith told the coroner that she had been William's wife for 'about twelve years' and that 'he had met with the accident *before* I married him'.[100] Although worthy of comment during the inquiry into his death, and clearly a physically restrictive feature of his life, William's industrial injury did not stop Barbara becoming his wife. As his court appearance for domestic violence in the 1860s indicates, Edward Rymer also managed to marry despite his impairments.[101] Indeed, Rymer's circumstances speak to a wider issue in coalfield society concerning the influence of impairment on a person's chances of marriage. With a 'damaged' right eye and 'scorched and frizzled' right side that left him 'permanently injured' following a fire at his home when he was around three years-old, Rymer had been impaired since early childhood.[102] This was something he shared in common with other mineworkers. Although few seem to have been so badly injured as early in life as Rymer was, the rigours of mining coupled with the young age at which miners tended to enter the industry meant many acquired impairments as children. Few mineworkers reached adulthood completely unscathed physically by their time in the mines. Impairment and chronic illness were so ubiquitous in mining communities that the prospect of a woman finding a husband free of permanent injury or ill health was slim indeed. Impairment in this context, then, was fairly unremarkable and women may have chosen to marry injured mineworkers because uninjured suitors were simply unavailable. However, given the positive associations of scarring and other bodily injuries as marks of bravery or experience in coalfield society, it is also possible that less incapacitating 'war wounds' may have added to a man's physical attractiveness.[103]

Religious life

While interpersonal relationships were tested and sometimes re-drawn by disability, the family was the bedrock of material and emotional support for sick, injured and impaired mineworkers. However, religious faith and fortitude and support from the spiritual community of believers were also important resources for some impaired coalminers. The final section of this chapter explores the place of religion in the lives of disabled mineworkers and their families.

A detailed study of the religious composition of mining communities goes beyond the scope of this study. However, historians have drawn attention to the strength and appeal of evangelical nonconformity, the various forms of Methodism in particular, in English and Welsh coalfields. At the vanguard of the evangelical revival sweeping across industrialising Britain in the eighteenth and early nineteenth centuries, Methodism fulfilled many needs in mining communities that were subject to economic upheavals, accidents and loss of life. As Colls, Moore and others have pointed out, Methodism helped explain suffering and death and offered solace to survivors, provided flexible educational opportunities for the young in communities that placed a premium on boys entering work in the mines from an early age, gave importance to lives otherwise worn down by repetitive and laborious work through involvement in class meetings, revivals and individual conversions, and provided roles for women beyond the walls of the domestic setting.[104] The appeal and durability of Methodism in the English and Welsh coalfields owes much to this broader social and community role, which was inextricably linked with its religious function.[105] Remembering his nineteenth-century ancestors, the Yorkshire miner Jim Bullock recalled that '[f]aith sustained them in sickness and poverty when everything seemed hopeless. The fact that they could tell God their troubles, and firmly believed that God did hear them and did something about it made it possible for them to bear life's hardships with patience and courage.' The importance of the chapel, he concluded, 'can never be over-valued'.[106]

As with other aspects of community life in the coalfields, however, we should be cautious about assuming homogeneity of experience. While the chapels of Methodists and other nonconformist sects provided 'people the opportunity to gain confidence in themselves', not everyone chose to avail themselves of these opportunities.[107] The chapel vied with other forms of leisure activity in pit villages.[108] The converted frequently saw themselves as engaged in a spiritual battle with non-believers who mocked or scoffed at their piety and while larger settlements might boast a rich variety of chapels and nonconformist sects, they were also home to many who lacked the time or inclination to attend divine service.[109] An account of the Hartley Colliery disaster of 1862 published in *The Wesleyan-Methodist Magazine*, for example, noted that although Methodism had 'done much to improve the habits and habitations of northern pitmen' there were few places where it had had *less* effect than in the community affected by the catastrophe, where fewer than one in ten employees of the pit were church members.[110] In some parts of the Scottish coalfield, sectarianism produced significant divides between mining households, leading as we saw in the previous chapter, to separate friendly societies to address the welfare needs of Protestants and Catholics.

While the role of religion in working-class politics during this period has been the subject of much historical debate, the place of religion in the lives of disabled workers and their families has received very little attention.[111] Religious leaders of all denominations played an active role in the aftermath of serious accidents. The visits of ecclesiastical worthies to pit communities following disasters in north-east England were frequently noted in the press.[112] After the Hartley disaster of 1862 it was reported that '[t]he Bishop of Durham, and other Christian pastors, with devout men and "ministering women" have visited the fatherless and widows in their affliction.'[113] Despite the strength of nonconformity in the north-east coalfield, the Anglican clergy conducted funerals, ministered to the bereaved and administered disaster relief funds.[114] Methodists too used gatherings to raise funds for widows and orphans. Following the explosion of fire damp at Jarrow Colliery in January 1826, a special sermon was given at the Methodist chapel in New Brunswick Place, Newcastle, 'in aid of the fund for the relief of the eleven widows, forty-six children, and several other dependant relations, of the unfortunate persons who lost their lives on the melancholy occasion'.[115] Local preachers provided spiritual comfort and leadership in local communities after accidents of all kinds.[116] In *Black Diamonds*, an anonymous account of missionary work in the south Staffordshire coalfield published in 1861, the author described spending a solid three or four weeks in 'domiciliary visitation' to the sick and injured of the thirty pits in the area.[117] For the faithful, belief in the power of prayer was just as important as – if not more than – medicine in recovery from accidents. The miner turned Methodist preacher, William Crister, for example, spent much time attending and praying for Mr Reay, his class leader, after he had 'met with a severe accident in the pit, by a fall of stone from the roof', from which his 'medical attendants' feared there was little hope of recovery. 'Till about the eighth or tenth day' after the accident, 'all were in despair except Crister' whose 'faith and hope' led him to the firm (and ultimately accurate) belief that Reay would survive.[118]

Placing particular emphasis on the importance of conversion, the authority of the Bible in providing lessons for faith and practice, and activism both in social causes and spreading the gospel, evangelical theology shaped thinking about accidents and disability in the religious culture of many mining communities.[119] The tomb-like darkness of the mine, its dangers and physical hardships, made the colliery a richly symbolic emblem of the toils and torments awaiting sinners in the 'future residence of bad men'.[120] Yet mining accidents also provided opportunities for examining other key aspects of religious faith, including fortitude, patience, assurance, providential deliverance and an opportunity to reflect on God's motives in taking or sparing life. Surviving a mining accident, for example, was sometimes cast as a seismic event in a collier's life that triggered conversion or intensification of religiosity.[121] These themes found expression in cheap religious tracts set in and around coalmining districts in the autobiographies of the converted and in a growing nonconformist newspaper and periodical press that was read widely in the coalfields.[122] The *Morning Chronicle*'s correspondent noted in 1849 that although the 'stock of books' was 'generally very small' in Durham pitmen's cottages, many had a 'large folio Bible' and a 'few Methodist tracts'.[123] Ironworkers and colliers residing in Merthyr's Cyfarthfa Row around the same time owned copies of exemplary conversion texts such as John Bunyan's *Pilgrim's Progress*.[124] Conversion narratives, 'a sort of Pilgrim's Progress kind of book' and 'books of a religious kind' in general, were also the most widely read texts among those employees of the Marquis of Londonderry's Durham collieries who attended the reading rooms provided in their villages, according to colliery supervisor G. Elliott in 1854.[125] Periodicals such as the *Wesleyan-Methodist Magazine* had a national circulation of 24,000 by the 1840s and contained many exemplary biographies of converted colliers.[126] The *Primitive Methodist Magazine*, speaking to the strongest branch of Methodism among the Durham miners, documented missionary work in the coalfields and printed moral tales.[127] Taken together, these publications offer insights into the relationship between accidents, disability and religious faith.

Given the importance of chronic illness and disablement in coalmining, religious texts frequently emphasised the importance of patience and fortitude in the face of life-changing conditions. Sufferings of the body always had spiritual significance.[128] *The Lancashire Collier Girl*, a cheap tract published in 1795, told the story of Mary, a young drawer in a colliery who witnessed the death of her father in an underground accident. Forced into the role of breadwinner after her grief-stricken mother 'became disordered in her senses', she continued to labour in the pit until she herself succumbed to a disabling illness that sapped her strength. In this narrative, a person's misfortunes and bodily afflictions were cast as a 'means of encreasing [sic] their trust in God'. Mary's faith was rewarded with being offered a place in domestic service by a local gentleman, leading the author to conclude that her tale showed that people should experience their 'afflictions' with forbearance.[129] The theme was reiterated in spiritual autobiographies and the evangelical press. An obituary of Thomas Cowey of Great Lumley, who converted to Methodism aged 16 after hearing Wesleyan ministers preaching at Shiney Row near Sunderland, related that he faced the many 'infirmities' occasioned by his work in the collieries with 'no murmuring' and was often heard to declare that '[a]fflictions are often God's choicest blessings'.[130] Durham miner George Parkinson related in his autobiography the story of 'Old Joe', one of his workmates who was also a Methodist lay preacher, who struggled to support his family during a nineteen-week lay off with a broken leg. With his wife, Sallie, unable to sleep for worry about their lack of money, Joe prayed for help and the next day discovered that a mysterious benefactor had paid for two weeks' groceries for the family. Bodily afflictions were a source of spiritual as well as material trial and ultimately this providential tale provided assurance of God's salvation for the converted.[131]

One of the most spiritually and emotionally testing experiences for mineworkers and others was amputation. Dismemberment was, as *The Wesleyan-Methodist Magazine* described it in 1836, a special 'trial' of 'faith and fortitude', which required considerable 'moral courage' on the part of patients.[132] For some, religious faith provided a source of strength with which to challenge the advice of surgeons to have damaged limbs removed. Following the accident in which his hand was crushed, Richard Weaver experienced 'pain' and 'fear of being unable now to earn his living', but 'was made to feel that it was all for the best' after 'God spoke to his heart' to say 'I will never leave thee, nor forsake thee.' Armed with this faith, Weaver refused the advice of the surgeon at Manchester Infirmary to have his hand removed in spite of medical opinion that the inflammation of his injury would kill him. 'If I die', Weaver reportedly told the surgeon, 'heaven will be my home; I don't fear death,' adding that 'Christ has taken away the fear of death, and I shan't let you take my hand off.' Weaver's faith was rewarded and over time 'in answer to prayer the inflammation subsided and it began to heal', although it remained 'permanently and seriously injured'.[133] Weaver's narrative simultaneously emphasised the frailty of the flesh and the power of faith in the hour of adversity. His accident and its aftermath was an important rite of passage both spiritually and physically. The incident took place not long after Weaver's conversion to Methodism and his joining the Wesleyan Society at Openshaw, and this was a crucial opportunity to test his faith and find assurance in God's protection. Although he was left with a lasting bodily impairment, he emerged from hospital spiritually empowered and ready to preach the gospel himself.[134]

The author of *Black Diamonds* (1861) similarly described amputation as a moment of spiritual significance in the case of 'George', an injured Staffordshire miner, who was 'regular in attending the evening school' that the author taught. George's thigh was 'fearfully crushed' after a horse-drawn skip was overturned onto it and he was sent to a 'neighbouring hospital'. The author described how he visited the hospital, following a request by George and his mother, and found that '[a]lthough his sufferings were of the most excruciating character, he was very pleased to see me, and was anxious to hear me speak about the Saviour, and to read the Scriptures, and pray with him,' such as to give the author hope that 'the work of grace had been commenced'. It fell to the author as a minister to convey to George the difficult advice of the house surgeon that his chances of survival would be greater if he had his severely infected leg removed. In spite of assurances that the use of chloroform would mitigate the pain of the operation, George was reluctant to undergo dismemberment, saying that without his leg 'I shall not be able to get my living.' Not wishing to live in a permanently disabled state, he declared that he 'should like to go to my Jesus at once'. Like Weaver, he 'appeared so happy in the assurance he had of pardon through the blood of Jesus, that he was not quite willing to have the dead limb amputated'. Nevertheless, 'after a little persuasion', George's wish to survive to see his chapel once more finally convinced him to have his limb removed. Three weeks after the operation, in spite of receiving the 'best attention' in hospital, he died.[135]

For those like George who dreaded the prospect of disability for its threat to a man's ability to earn a living and its risk of dependency on others, religious faith could provide a source of strength and self-determination. Those who were unable to work were represented as facing many temptations and needed inner strength to avoid falling into despondency. If the heavy labours of the mineworker demonstrated the physical strength and bravery of the 'outer' man, the spiritual struggles of the incapacitated helped assert men's 'inner' qualities. Although physical infirmities sapped the strength of Durham miner Thomas Cowey, his friends 'rejoiced to see the proofs of increasing strength in the inner man' as he bore with his 'afflictions' stoically and through faith achieved 'inward happiness'.[136] An obituary of Joseph Wailes, a Cumberland lead miner, published in the *Primitive Methodist Magazine* in 1843 described how he, like many underground workers, had been confined to his room by a 'pulmonary ailment which settled in a consumption'. During this confinement he was 'powerfully assailed with Satan' and was tempted with the idea that his Christian religion was 'delusive', but thanks to the power of prayer and the support of his fellow converts, he 'gained a complete conquest' over his adversary. These struggles with Satan in the sickroom provided edifying examples of how disease, incapacity and confinement might ultimately result in spiritual triumph, even if the physical body was defeated and beyond repair.[137]

One of the hardest consequences of physical incapacity for patients to bear was that in an evangelical religious culture that valued sharing religious experience through Bible reading, class meetings, revivalist gatherings and chapel membership, it potentially robbed converts of the company of the faithful.[138] During a lengthy 'affliction' with 'severe and prostrating' symptoms that confined him to his bed, William Lee of Netherton Colliery, near Morpeth, was 'deprived ... of the blessings of public worship, and deeply felt the loss of the communion of saints, and of the ministry of the Word of Life'.[139] However, the community activism that characterised evangelical nonconformity might provide some opportunities for people whose impairments were less restrictive, as several exemplary cases reported in the Methodist press revealed. The *Methodist Magazine* in 1805 related the life of Thomas Handley of Coalbrookdale in Shropshire who, after losing one of his legs, was 'stirred up to seek the Lord in earnest' and visited local villages 'to warn the thoughtless inhabitants to flee the wrath to come'.[140] As we saw earlier, Richard Weaver's ministry began after the accident that left his hand permanently incapacitated.

The community activism of women with disabilities was also celebrated in the Methodist press. An account of Margaret Crozier of Pelton Fell, published in 1875, showed how faith and good works gave meaning to the life of a disabled wife of a Durham pitman. Margaret was 'for over thirty years a stranger to perfect bodily health' and had sustained a head injury after falling from a second storey window as a girl, 'which was the cause of intense suffering, physical and mental all through her after life'. She was widowed at age twenty-nine after her husband (a lay preacher and class leader) died in an explosion at Pelton Colliery. Left with three dependent children and suffering from a 'frail physical constitution', she embarked on a life of exemplary piety, both as a mother and as a 'most neighbourly woman', who undertook many 'weary watchings by the sick and dying'. Though this readiness to provide solace to the sick, injured and bereaved of her mining village 'involved great sacrifice and inconvenience', all was 'done and borne with a Christian cheerfulness' that earned her the respect of her community.[141]

Conclusion

This chapter has shown that injury, impairment and chronic illness had significant consequences for miners' lives beyond the workplace. They might affect mineworkers' emotional lives, family relationships and standing in the social and spiritual life of coalfield communities. The effects of disablement were gendered. Women as well as men experienced impairment in Britain's coalfield communities both through paid work (particularly before females were banned from underground labour in 1842) and through injury and ill health occasioned by childbirth and the arduous unpaid work of washing, cleaning and carrying water that was essential to servicing the industry. However, such impairments were far less publicly visible than those of male breadwinners. In a strongly masculine culture that valorised strength and the ability to provide, the impaired miner faced an uncertain future and potentially had to negotiate a loss of status in the community, at home as well as at work. However, emasculation was not inevitable and there were ways in which impaired miners might seek to regain control and shape meaningful lives that rebuilt their masculine identity, both in their own eyes and the eyes of others.

These strategies varied. The older stereotype of the miner as tough, hard-drinking and prone to violence, though losing some of its force as the period progressed, provided a means by which some impaired colliers sought to assert themselves, using fighting as a means of 'proving' their strength in the face of serious injury. Such forms of expression stood in contrast with Victorian ideals of temperance and respectability that emphasised self-control and restraint. However, the recourse to violence on the part of some impaired miners, both on the streets and in the home, shows its continued importance as a ready means of asserting potency. Others sought roles in their communities that earned respect through setting an example to others rather than through fear. The spread of evangelical nonconformity, with its emphasis on communal solidarity of the converted and social activism, offered new roles for impaired men and women in coal communities. For some miners facing a re-evaluation of their lives and prospects in the face of serious injury or disease, personal faith and involvement in the religious lives of their community provided means of coping with change and new forms of empowerment. While many of the examples of religious strength that survive in autobiographies or obituaries of the faithful printed in the Methodist press were often idealised and exemplary in function, they nonetheless highlight the significance of evangelical religion as a tool for adapting to impairment in industrialising Britain.

It was within the family that the effects of impairment or chronic illness were felt most keenly. The ideals of female domesticity and male breadwinning were tested by impairment, which disrupted and re-drew conventional relationships within the home. However, while men's patriarchal authority was challenged by impairment, it often remained resilient in the face of physical adversity. In some cases, men were able (and were expected) to continue to make some provision for their families, even though the balance of economic power might shift towards their children. The continued retaining of the 'headship' role within families by impaired men – symbolically at least – helped them to maintain status within their communities and to mitigate the demeaning prospect of dependency.[142] And while unmarried men might see their potential worth as suitors diminish after impairment, their marriage prospects were rarely damaged irreparably.

Mining communities were built on a cultural ideal of solidarity in the face of shared exposure to danger. Disabled people were supported, but they might also face distrust – especially strangers or those whose impairments drew suspicion. Differing approaches to disability in coalfield communities were exposed in times of political tension and unrest. The final chapter of this book turns its attention to the place of disability in the industrial politics of coalmining in nineteenth-century Britain.

Notes

1 Jelinger C. Symons, *Tactics for the Times: As Regards the Condition and Treatment of the Dangerous Classes* (London: John Olliver, 1849), 32–3.
2 P. E. Razzell and R. W. Wainwright (eds), *The Victorian Working Class: Selections from Letters to the Morning Chronicle* (London: Frank Cass, 1973), 230.
3 Symons, *Tactics*, 33–4.
4 K. S. Inglis, *Churches and the Working Class in Victorian England* (London: Routledge and Kegan Paul, 1963), 10, cited in Colin P. Griffin, 'Methodism in the Leicestershire and South Derbyshire Coalfield in the Nineteenth Century', *Proceedings of the Wesley Historical Society*, 39 (1973–4), 64.
5 R. C. Morgan, *The Life of Richard Weaver, the Converted Collier* (London: Morgan and Chase, 1861), 98–9.
6 John Benson, 'The Thrift of English Coal-Miners, 1860–95', *Economic History Review*, 31:3 (1978), 410–18.
7 'The Collier at Home', *Household Words*, 15:366, 28 March 1857, 291.
8 Ibid., 290, 291.
9 Jules Ginswick (ed.), *Labour and the Poor in England and Wales 1849–1851: the Letters to the Morning Chronicle from Correspondents in the Manufacturing and Mining Districts, the towns of Liverpool and Birmingham, and the Rural Districts*, vol. 3: *South Wales–North Wales* (London: Frank Cass, 1983), 49.
10 For example, John Benson *British Coalminers in the Nineteenth Century: A Social History* (Dublin: Gill and Macmillan, 1980); Robert Colls, *Pitmen of*

the Northern Coalfield: Work, Culture, and Protest, 1790–1850 (Manchester: Manchester University Press, 1987); Robert Colls, *The Collier's Rant: Song and Culture in the Industrial Village* (London: Croom Helm, 1977); R. I. Moore, *Pit-Men, Preachers and Politics: the Effects of Methodism in a Durham Mining Community* (Cambridge: Cambridge University Press, 1974); David Gilbert, *Class, Community and Collective Action: Social Change in Two British Coalfields, 1850–1926* (Oxford: Oxford University Press, 1992); Alan Metcalfe, *Leisure and Recreation in a Victorian Mining Community: the Social Economy of Leisure in North-East England, 1820–1914* (London: Routledge, 2006); Jaclyn J. Gier and Laurie Mercier, *Mining Women: Gender in the Development of a Global Industry* (New York: Palgrave Macmillan, 2006).
11   Moore, *Pit-Men*, 16; Benson, *British Coalminers*, 82–4.
12   Metcalfe, *Leisure*, 16.
13   Paul O'Leary, *Immigration and Integration: The Irish in Wales 1798–1922* (Cardiff: University of Wales, 2000), 44–5; Alan B. Campbell, *The Lanarkshire Miners: a Social History of their Trade Unions, 1775–1874* (Edinburgh: John Donald, 1979).
14   Metcalfe, *Leisure*, 21–3; David Gilbert, 'Imagined Communities and Mining Communities', *Labour History Review*, 60:2 (1995), 47–55.
15   [Edward Chicken], *The Comical History of the Collier's Wedding in Fyfe. Wrote by a Kirkaldy Gentleman* (Edinburgh, 1779), 3.
16   For the fullest account of changes in popular cultural images of coalminers, see Colls, *Collier's Rant*.
17   Thomas Wilson, *The Pitman's Pay and other Poems* (Gateshead: William Douglas, 1842), vii–viii.
18   'The Miners Life', *The Motherwell Times*, 4 April 1885.
19   Bruce Haley, *The Healthy Body and Victorian Culture* (Cambridge, MA and London: Harvard University Press, 1978); Jennifer Esmail and Christopher Keep, 'Victorian Disability: Introduction', *Victorian Review*, 35:2 (2009), 45–51.
20   *British Miner and General Newsman: A Publication Devoted to the Interests of the Working Miners of the United Kingdom*, 4, 4 October 1862, 5.
21   Ibid.; Colls, *Pitmen*, 122–45.
22   'Assaulting the Police', *The Merthyr Telegraph*, 23 August 1862.
23   *The Merthyr Telegraph*, 28 November 1863; ibid., 25 April 1873. 'Mockyn' seems to be a version of mochyn, Welsh for pig, whereas 'croes-pen' could translate as 'cross head', suggesting someone prone to anger. We are grateful to Carys Turner for advice on translation.
24   Andy Croll, *Civilizing the Urban: Popular Culture and Public Space in Merthyr, c. 1870–1914* (Cardiff: University of Wales Press, 2000).
25   *Glasgow Herald*, 16 February 1875.
26   Croll, *Civilizing the Urban*; W. R. Lambert, *Drink and Sobriety in Victorian Wales, c. 1820–1895* (Cardiff: University of Wales Press, 1983); Brian Harrison, *Drink and the Victorians: the Temperance Question in England, 1815–1827* (London:

Faber, 1971); Julie Marie Strange, *Fatherhood and the British Working Class, 1865–1914* (Cambridge: Cambridge University Press, 2015).

27 Shani D'Cruze, *Crimes of Outrage: Sex, Violence and Victorian Working Women* (DeKalb: Northern Illinois University Press, 1998), 115.

28 For example, Lenore Manderson and Susan Peake, 'Men in Motion: Disability and the Performance of Masculinity', in C. Sandahl and P. Auslander (eds), *Bodies in Commotion: Disability and Performance* (Ann Arbor: University of Michigan Press, 2005), 230–42.

29 D'Cruze, *Crimes of Outrage*, 140.

30 'A Dangerous Leg', *The Merthyr Telegraph*, 29 September 1866. Our emphasis.

31 James Gregory, 'Eccentric Lives: Character, Characters and Curiosities in Britain c. 1760–1900', in Waltraud Ernst (ed.), *Histories of the Normal and the Abnormal: Social and Cultural Histories of Norms and Normativity* (London and New York: Routledge, 2006), 73–100. See also Iain Hutchison, *A History of Disability in Nineteenth-Century Scotland* (Lewiston, NY: Edwin Mellen Press, 2007), 61.

32 'The Bard of Pen-Heol-Gerrig', *The Merthyr Telegraph*, 29 May 1858.

33 James Dalglish, 'Explosion of Gas in a Coal-Mine', *Methodist Magazine*, 5 (April 1826), 242.

34 Such scenes were a popular theme for visual representation in the illustrated press. For example, see 'Funeral of the Colliers Killed by the Late Explosion at Tredegar', *Illustrated London News*, 1 July 1865; 'Dinas Colliery, Rhondda Valley, South Wales, the Scene of the Late disaster', *Illustrated London News*, 23 January 1879.

35 Jim Bullock, *Bowers Row: Recollections of a Mining Village* (Wakefield: E. P. Publishing, 1976), 220.

36 'Hunting a Packman', *The Aberdare Times*, 6 May 1865.

37 Keith Strange, *Merthyr Tydfil Iron Metropolis: Life in a Welsh Industrial Town* (Stroud: Tempus, 2005), 67.

38 David M. Turner, '"Fraudulent" Disability in Historical Perspective', *History and Policy* (14 February 2012), http://www.historyandpolicy.org/policy-papers/papers/fraudulent-disability-in-historical-perspective, accessed 27 June 2016.

39 Ginswick (ed.), *Labour and the Poor* ... vol. 3, 67.

40 O'Leary, *Immigration and Integration*, 186–212.

41 John Wilson, *Autobiography of Ald. John Wilson, J.P., M.P ... Reprinted from 'The Durham Chronicle'* (Durham: Durham Chronicle, 1909), 19.

42 *Monmouthshire Merlin*, 17 May 1862. For a similar case see 'A Begging Imposter', *The Cardiff and Merthyr Guardian*, 14 May 1870.

43 Edward Rymer, *The Martyrdom of the Mine, or a Sixty Years Struggle for Life* (Middlesbrough, 1898), 14.

44 Chris Fisher and Pat Spaven, 'Edward Rymer and "The Moral Workman" – The Dilemma of the Radical Miner Under "MacDonaldism"', in Royden Harrison (ed.), *Independent Collier: the Coal Miner as Archetypal Proletarian Reconsidered* (Hassocks: Harvester Press, 1978), 232–71.

45 Rymer, *Martyrdom of the Mine*, 8; 'Providence against the Tories', *Dean Forest Mercury*, 13 November 1885.
46 James Dunn, *From Coal Mine Upwards, Or, Seventy Years of an Eventful Life* (London: W. Green, 1910), 1–19, quote from 4.
47 For further critique of the solidarity of coalfield communities, see Gilbert, 'Imagined Communities and Mining Communities'.
48 'The Strike in South Wales: Interior of a Collier's Cottage', *Illustrated London News*, 18 January 1873.
49 William Stuart Howard, 'Miner's Autobiographies, 1790–1945: A Study of Life Accounts by English Miners and their Families' (unpublished PhD thesis, Sunderland Polytechnic, 1991), 170–204.
50 Jane Humphries, *Childhood and Child Labour in the British Industrial Revolution*. (Cambridge: Cambridge University Press, 2010), 151; Kathryn Gleadle, *British Women in the Nineteenth Century* (New York: Palgrave Macmillan, 2001), 188.
51 Carluke, Lanarkshire, 1871 Scotland Census.
52 University of Glasgow Mitchell Library, Glasgow City Archives, CO1 27 91 Parish of Carluke, Record of Application for Relief, Years 1866–77.
53 Carluke, Lanarkshire, 1871 Scotland Census; University of Glasgow Mitchell Library, Glasgow City Archives, CO1 27 91 Parish of Carluke, Record of Application for Relief, Years 1866–77.
54 'Letter to Her Majesty the Queen on the Relief of the Sufferers by the Late Gethin Explosion', *Merthyr Telegraph*, 20 January 1866.
55 Humphries, *Childhood and Child Labour*, 71–2.
56 Carluke, Lanarkshire, 1871 Scotland Census.
57 Cf. PP 1842 (008), *Sanitary Enquiry – Scotland, Reports on the Sanitary Condition of the Labouring Population of Scotland, in Consequence of an Inquiry Directed to be Made by the Poor Law Commissioners*, 98; John R. Gillis, *For Better, for Worse: British Marriages, 1600 to the Present* (New York and Oxford: Oxford University Press, 1985), 200–1.
58 PP 1842 (008), 98; J. R. Leifchild, *Our Coal and our Coal Pits* ([1856] London: Frank Cass, 1968), 197.
59 Michael R. Haines, 'Fertility, Nuptiality, and Occupation: a Study of Coal Mining Populations and Regions in England and Wales in the Mid-Nineteenth Century', *Journal of Interdisciplinary History*, 8 (1977), 267; cf. Mary Jo Maynes, 'Fertility and Occupation: Population Patterns in Industrialization by Michael Haines' (review), *Journal of Interdisciplinary History*, 11 (1981), 712–14; Simon Szreter, *Fertility, Class and Gender in Britain, 1860–1940* (Cambridge: Cambridge University Press, 1996), 526–7.
60 Haines, 'Fertility, Nuptiality, and Occupation'; Humphries, *Childhood and Child Labour*, 55–7; Gleadle, *British Women*, 123–9; Emma Griffin, *Liberty's Dawn: A People's History of the Industrial Revolution* (New Haven, CT and London: Yale University Press, 2013), 107–62.
61 PP 1842 (381), 451.

62 Humphries, *Childhood and Child Labour*, 254.
63 Razzell and Wainwright (eds), *Victorian Working Class*, 226; M. J. Daunton, 'Miners' Houses: South Wales and the Great Northern Coalfield, 1880–1914', *International Review of Social History*, 25:2 (1980), 153.
64 'Letter to Her Majesty the Queen on the Relief of the Sufferers by the Late Gethin Explosion', *Merthyr Telegraph*, 20 January 1866.
65 Ginswick (ed.), *Labour and the Poor* ... vol. 3, 54; Metcalfe, *Leisure*, 18.
66 Andrew Walker, '"Pleasurable Homes"? Victorian Model Miners' Wives and the Family Wage in a South Yorkshire Colliery District', *Women's History Review* 6 (1997), 327.
67 Razzell and Wainwright (eds), *Victorian Working Class*, 252.
68 Leifchild, *Our Coal*, 197–198. For earlier comments of this nature that blamed working in mines for women's lack of time for maternal duties, see Robert Bald, *A General View of the Coal Trade of Scotland* (Edinburgh: Oliphant, Waugh and Innes, 1812), 138.
69 Howard, 'Miner's Autobiographies', 170–1.
70 Dot Jones, 'Counting the Cost of Coal: Women's Lives in the Rhondda, 1881–1911' in Angela V. John (ed.), *Our Mothers' Land: Chapters in Welsh Women's History, 1830–1939* (Cardiff: University of Wales Press, 1991), 124–6. See also Colls, *Pitmen*, 257–8.
71 *The Cambrian*, 13 April 1839; *The Glamorgan, Monmouth and Brecon Gazette and Merthyr Guardian*, 6 April 1839.
72 *Glasgow Herald*, 31 October 1855.
73 Humphries, *Childhood and Child Labour*, 43, 237, 368–70.
74 'Births in Colliers' Families', *Paisley Herald and Renfrewshire Advertiser*, 22 December 1855.
75 John Saunders, *Israel Mort, Overman* (London: Henry S. King, 1876), 18, 182. See also Alexandra Jones, 'Disability in Coalfields Literature c. 1880–1948: A Comparative Study' (unpublished PhD thesis, Swansea University, 2016), 49–50.
76 Julie Marie Strange, *Fatherhood and the British Working Class, 1865–1914* (Cambridge: Cambridge University Press, 2015), 51.
77 Burt, *Autobiography*, 134–5.
78 PP 1842 (008), 99.
79 *Durham County Advertiser*, 8 February 1861.
80 Humphries, *Childhood and Child Labour*, 135.
81 Strange, *Fatherhood*, 54.
82 Benson, 'Thrift of English Coalminers', 412.
83 *The Poor Law Unions' Gazette*, 14 November 1857.
84 Robert Bald, *A General View of the Coal Trade of Scotland, Chiefly that of the River Forth and Mid-Lothian* (Edinburgh: Oliphant, Waugh and Innes, 1812), 138.
85 PP 1842 (381), 475.
86 As Bourke has pointed out, working-class women were often reluctant to enter paid employment because they realised it would significantly increase their

workload. [Joanna Bourke, 'Housewifery in Working-Class England 1860–1914', *Past & Present*, 143 (1994), 173]. Given this, Boxter's example also suggests that disability could threaten the health of other family members, particularly women. In other words, by disturbing the gendered organisation of mining households, but not destroying it, disability may have increased pressure on women like Boxter to work even more than other working-class women did, thereby further undermining their health in the ways Dot Jones has argued.

87 For more on disability and the 'performance' of gendered identities, see Manderson and Peake, 'Men in Motion'.
88 Howard, 'Miner's Autobiographies', 173; Colls, *Collier's Rant*, 119.
89 For a sense of the breadth and scope of this work, see Robert McRuer and Anna Mollow (eds), *Sex and Disability* (Durham, NC: Duke University Press, 2012).
90 PP 1842 (381), 475.
91 For example, Rymer, *Martyrdom of the Mine*, 16.
92 Carluke, Lanarkshire, 1871 Scotland Census.
93 Humphries, *Childhood and Child Labour*, 53.
94 Ava Baron, 'Masculinity, the Embodied Male Worker, and the Historian's Gaze', *International Labor and Working-Class History*, 69 (2006), 151. On 'headship' and disabled men's status see Daniel Blackie, 'Disability, Dependency, and the Family in the Early United States' in Susan Burch and Michael Rembis (eds), *Disability Histories* (Urbana IL: University of Illinois Press, 2014), 17–34.
95 For an in-depth consideration of the many variables affecting the experiences of disability in Welsh mining families during a later period, see: Ben Curtis and Steven Thompson, 'Disability and the Family in South Wales Coalfield Society, c. 1920–1939', *Family & Community History* 20:1 (2017), 25–44.
96 Weaver's accident and its consequences for his spiritual development are discussed further below.
97 Richard Weaver, *Richard Weaver's Life Story: the English Evangelist*, ed. James Patterson (Kilmarnock: J. Ritchie, 1897), 61.
98 Bourke, 'Housewifery', 196.
99 PP 1833 (450), *Factories Inquiry Commission. First Report ... with Minutes of Evidence, and Reports by the District Commissioners*, minutes of evidence taken by Mr Tufnell, D (Lancashire District), 79.
100 *Cardiff and Merthyr Guardian*, 12 June 1847. Our emphasis.
101 *Durham County Advertiser*, 8 February 1861; Rymer, *Martyrdom of the Mine*, 7.
102 Ibid., 2.
103 Edward Slavishack, *Bodies of Work: Civic Display and Labor in Industrial Pittsburgh* (Durham NC and London: Duke University Press, 2008), 162.
104 Colls, *Pitmen*, part 2 *passim*.; Moore, *Pit-Men, Preachers and Politics*; Hempton, *Methodism and Politics*, 214.
105 Griffin, 'Methodism in the Leicestershire and South Derbyshire Coalfield', 71.
106 Bullock, *Bowers Row*, 146
107 Ibid., 141; Griffin, *Liberty's Dawn*, 201.

108 Metcalfe, *Leisure*, ch. 2.
109 Strange, *Merthyr Tydfil Iron Metropolis*, 119.
110 Henry Hine, 'The New-Hartley Pit Catastrophe', *The Wesleyan-Methodist Magazine*, 8 (March 1862), 252.
111 E. P. Thompson, *The Making of the English Working Class* (Harmondsworth: Penguin, 1976), 350–400; Moore, *Pit-Men, Preachers and Politics*, 8–12; Hempton, *Methodism and Politics*, ch. 7.
112 Jonathan Fryer, 'Preachers at the Pit: Methodists and County Durham Mining Disasters, 1880–1909', *Proceedings of the Wesley Historical Society*, 57:2 (2009), 27.
113 Hine, 'New-Hartley Pit Catastrophe', 253.
114 Robert Lee, *The Church of England and the Durham Coalfield 1810–1926: Clergymen, Capitalists, and Colliers* (Woodbridge: Boydell Press, 2007), 254–7.
115 Dalglish, 'Explosion of Gas', 241.
116 Fryer, 'Preachers at the Pit', 27.
117 H. H. B., *Black Diamonds; or, the Gospel in a Colliery District* (London: James Herbert et al., 1861), 74.
118 James Everett, *The Wall's End Miner, or, A Brief Memoir of the Life of William Crister including an account of the Catastrophe of June 18th, 1835* (London: Hamilton, Adams and Co, 1835), 78. For later examples see Jones, 'Disability in Coalfields Literature', 135.
119 David Bebbington, *Evangelicalism in Modern Britain: A History from the 1730s to the 1980s* (London: Routledge, 1989), ch. 1.
120 John Crosby, 'Some account of the Experience of Nathaniel Harrison of Bingley-Moor, near Leeds', *Methodist Magazine* (December 1801), 526.
121 For example, 'Obituaries' (Ralph Hall of Gateshead Circuit d. 3 January 1822), *Wesleyan-Methodist Magazine* 1 (April 1822), 272; E. H., 'Recent Deaths' (William Lee, Netherton Colliery, d. 17 February 1871), *The Wesleyan-Methodist Magazine*, 18 (May 1872), 478; J. Stephenson, 'Extraordinary Delivery of Two Colliers', *Primitive Methodist Magazine*, 3rd ser., vol. 1 (1843), 243. The theme remained popular in Methodist coalfield literature until after the First World War: Jones, 'Disability in Coalfields Literature', 126–32.
122 See, for example, J. A. James, *The Pious Collier, or the Life of Joseph Round* (London: Religious Tract Society n.d.); *Perils in the Mine. A Colliery Tale, In Verse* (London: Jarrold and Sons, n.d. [c. 1863]).
123 Razzell and Wainwright (eds), *Victorian Working Class*, 227.
124 Ginswick (ed.), *Labour and the Poor* ... vol. 3, 53.
125 PP 1854 (258), *Second Report from the Select Committee on Accidents in Coal Mines; with the Minutes of Evidence Taken Before Them*, 18.
126 Hempton, *Methodism and Politics*, 182.
127 Colls, *Colliers' Rant*, 84.
128 Stuart Hogarth, 'Joseph Townend and the Manchester Infirmary: a Plebeian Patient in the Industrial Revolution' in Anne Borsay and Peter Shapely (eds),

Medicine, Charity and Mutual Aid: the Consumption of Health and Welfare in Britain c. 1550–1950 (Aldershot: Ashgate, 2007), 103.
129 Anon., *The Lancashire Collier Girl. A True Story* (London: J. Marshall, 1795), 7, 15, 21–2.
130 James Bromley, 'Brief Biographical Sketches', *The Wesleyan-Methodist Magazine*, 3 (December 1847), 1155.
131 George Parkinson, *True Stories of Durham Pit-Life* (London: C. H. Kelly, 1912), 36–7.
132 Thomas Beaumont, 'Memoir of Miss Grace Elizabeth Wilkinson, of Bradford, Yorkshire', *The Wesleyan-Methodist Magazine*, 15 (February 1836), 97.
133 Morgan, *Life of Richard Weaver*, 25–6. See also Weaver, *Richard Weaver's Life Story*, 60–1.
134 Weaver's case is similar to that of Manchester Methodist Joseph Townend: Hogarth, 'Joseph Townend and the Manchester Infirmary', 103–5.
135 H. H. B., *Black Diamonds*, 193–6.
136 Bromley, 'Brief Biographical Sketches', 1155–6.
137 'Obituary', *Primitive Methodist Magazine*, Third Series, 1 (1843), 359.
138 Griffin, *Liberty's Dawn*, 193.
139 E. H., 'Recent Deaths', 479.
140 Samuel Taylor, 'The Grace of God Manifested, in a short Account of the Conversion, Experience and Happy Death of Thomas Handley, and Several Others', *Methodist Magazine*, 28 (November 1805), 505–8.
141 Forster Crozier, 'Memoir of Mrs. Margaret Crozier, of Pelton Fell, Durham', *The Wesleyan-Methodist Magazine*, 21 (April 1875), 295–301.
142 Blackie, 'Disability, Dependency, and the Family', 25.

5

THE INDUSTRIAL POLITICS OF DISABLEMENT

In 1843, one year after Parliament had passed the landmark Mines and Collieries Act banning females and children under the age of 10 from working underground, *Punch* magazine printed a cartoon titled 'Capital and Labour' (Figure 4). Reflecting the magazine's sympathy for the poor and downtrodden and the spirit of social justice that characterised its radical early years, the image contrasted the circumstances of those who grew rich from coalmining, the wealthy coal owners, with those whose work brought their profits. Beneath the soft opulence of the capitalist's dwellings is depicted a prison-like coal mine, populated by the vulnerable and impaired: starving ragged children, a frightened-looking woman holding a baby to her breast, an old man with twisted limbs bent double, another on crutches, another lying broken and exhausted on the ground. Watched over by a fat gaoler who collected the employer's 'gold', this was a world from which hope and love were firmly shut out – as shown by their allegorical figures, pictured to the left of the underground scene, pushing at a locked door.[1]

Punch's cartoon referenced the vivid images of working conditions included in the Children's Employment Commission Report, which had caused a press sensation on their publication in May 1842. The Report's depiction of the 'dismal chambers' in which men, women and children worked, in tasks that required 'the severest exertions' to complete, exposed the rigours of mine work to horrified legislators and members of the public.[2] Although it was not the first official publication to recognise the dangers of coalmining, the report represented a key moment in the emergence of the cultural figure of the disabled mineworker. This figure was used rhetorically to great effect by policymakers and critics of industrialisation during the frequent debates about mining regulation that punctuated the nineteenth century. In persuading Parliament to outlaw the underground employment of women and young children in 1842,

Figure 4 'Capital and Labour', *Punch*, 29 July 1843. Reproduced with the permission of Punch Ltd. Punch.co.uk.

the law's chief architect, Lord Ashley, drew heavily on the report's evidence of the injurious effects of mine work, citing statements about the 'fatigue', 'deformity', 'distorted' spines and other assorted injuries and diseases suffered by mine-working women and children.[3] Such references helped supporters of industrial reform win their argument in 1842 and testified to the power of the image of the disfigured and disabled coal worker in the popular imagination.

As we have seen, at one level the 1842 Mines and Collieries Act was a piece of moral legislation, intended to end the exploitation of very young children in collieries and also to reinforce Victorian domestic ideology. By ending female work underground, it was hoped the law would encourage women and girls to remain at home, where it was thought they belonged. Yet the 1842 Act was about much more than promoting gendered social roles. As the law stated, one of its main goals was to 'make Provisions for the safety' of mineworkers – by mandating, for example, age limits on tasks such as the operation of shaft engines.[4] The Act also empowered an inspector to enter a coal mine to check whether the provisions of legislation were 'properly observed' and report any breaches to the Home Office.[5] Although the appointment of only one inspector, Hugh Seymour Tremenheere, to monitor all mines in Britain was clearly inadequate, over the course of the following thirty years government regulation of coalmining was progressively extended.[6] This included the establishment in 1850 of a more comprehensive Mines Inspectorate with greater powers to fine coal owners for non-compliance with safety recommendations and (as we saw in Chapter 1) the better documentation of fatal accidents and 'serious injuries'.[7] Aimed as they were at reducing or documenting the number of accidents, such measures indicate policymakers' growing concerns about industrial death and injury in the nineteenth century and their willingness to do something about it.

As Catherine Mills has shown, the tightening of laws around mining safety and inspection owed much to high occupational mortality, the increasing visibility of risk, and developing medical knowledge about miners' susceptibility to illnesses such as lung disease. Legislative action was not only down to the goodwill of legislators', but also stimulated by a combination of public sympathy and trade union activism spurred on by the extension of the franchise.[8] Legal reforms also reflected shifting ideas in political economy. As we have seen, at the beginning of the nineteenth century it was widely believed that miners' relatively good wages compensated them for the higher risks they faced, which, taken together with the employer paternalism examined in previous chapters, negated the demand for government regulation to protect health and safety.[9] However, over the course of the century this was gradually replaced by the idea that workers were 'bodily capital' that needed protecting in the interests of national prosperity and competitiveness. Mine and factory regulation shifted over the course of the middle part of the nineteenth century away from a narrow and paternalistic concern with the health and moral well-being of women and children, towards interventions that sought to protect the health and productivity of all workers, including men.[10] While the prevention of fatalities remained a priority of policymakers, safety regulations that would reduce non-fatal injuries as well were introduced. By 1872, mines were required to have at least two shafts to ensure ventilation and emergency access, and managers were obliged to take measures to 'secure' roofs and fence off machinery.[11] Not only did these reforms try to prevent or reduce the number of accidents, they also came to define the responsibilities of the interests of 'Capital' to men and women of 'Labour' in compensating those injured in the workplace.

The history of mining regulation has been well documented, but few accounts have explored the place of *disability* in nineteenth-century coalfield industrial relations. As Sarah Rose points out, as valuable as state-centred accounts of disability policies and politics are, they tend to overlook the experiences of disabled people and the role they have played in shaping their own lives. Over the course of this period, miners and their representatives became increasingly vocal in drawing attention to the daily threats posed to their 'lives and limbs' in the coal mines. In its use of broken bodies to depict the injustices of capitalism, Punch's cartoon represented in visual form a rhetoric of disability that was increasingly used in the nineteenth century as a rallying call for men to combine in local or national trade unions. Disablement, as we shall see, was represented as a shared risk that bonded working people together in a common experience of suffering.[12] Exploring both local and national campaigns, this chapter examines the ways in which unions took up the cause of disabled miners and provided their own systems of support. It also examines the political agency of disabled miners themselves and explores their experiences of industrial conflict. Although we should not exaggerate the extent of trade union membership or political activism among Britain's coalminers before 1880, fluctuating relations between employers and mineworkers often affected the lived experiences of disabled people in dramatic ways.[13] For many miners, the industrial politics of disablement played out in local settings and was concerned not just with working conditions, but also how those left injured or permanently impaired were treated. Industrial relations raised questions of responsibility and loyalty, in which miners' expectations of their employers – and their fellow workmen – came under scrutiny.

This chapter asks how seriously labour leaders and employers took the needs of disabled workers. Both sides frequently proclaimed they had the best interests of disabled workers at heart. Nevertheless, nineteenth-century coal owners and mining unions were often prepared to sacrifice the immediate needs of sick and injured miners in pursuit of victory in the heated industrial conflicts in which they were frequently engaged. Disabled miners and their families – like inhabitants of mining communities more generally – undoubtedly faced increased hardships during strikes. However, periods of intense industrial conflict may not always have been unmitigated disasters for disabled people. On the contrary, this chapter presents evidence that industrial disputes may have presented disabled mineworkers with opportunities that some chose to exploit.

Disability and industrial relations: the voices of labour

The years following the repeal of the repressive Combination Laws of the Napoleonic era in 1824 witnessed an 'upsurge of trade union organisation', especially among colliers.[14] Miners' concerns about injury and illness were central to this organisational activity. In order to attract members, many mining unions included sick pay among the benefits they offered. For example, in the 1830s, Thomas Hepburn's Union of Miners in north-east England offered sickness benefits to those who joined. During the cholera epidemic of 1831–32, this commitment to members proved a serious drain on the union's financial resources and undermined its ability to maintain effective strike action. By mid-1832, the union was spending almost 40 per cent of its income on sickness and death benefits.[15]

As John Benson has shown, after 1850 a significant majority of English mining unions operated sickness and accident funds. These funds had three main purposes. First, they aimed to attract members by offering a benefit that would appeal to miners given the dangers of their occupation. Second, based as they were on mutualist principles, such benefits sought to promote solidarity and fraternal feelings among members. Third, sickness and accident pay was intended to reduce miners' dependence on employer-controlled schemes such as pit clubs. More than simply a pragmatic response to ill health or injury, then, union provisions for sick or injured members aimed to maintain the dignity, independence and self-determination of workers that was essential for the success of trade unionism.[16] Like other self-help schemes, however, trade unions were not very good at providing long-term support for permanently disabled members. In 1872, for example, the Durham Miners' Association promised five shillings a week to members incapacitated for work for the first six months of incapacity, falling to half that amount thereafter. After a year, the weekly benefit fell further still, to two shillings, 'so long as such a person be unable to resume work'.[17]

Mirroring friendly societies, trade union funds also often imposed moral clauses on their members that aimed to uphold the ideal of the careful, responsible worker. In Scotland, the Larkhall Miners' Mutual Protection, Accident and Funeral Association, founded in 1874, for example, emphasised 'mutual support', including making provision for 'members when disabled by accident in following their employment', but fined men who were found drunk at work or were deemed 'guilty of any grossly culpable act while on duty'. Worse still, offenders could lose their right to benefits for up to three months, depending on the will of the association's Board of Management.[18] Members who endangered their workmates, or were injured through their own irresponsibility, then, were stigmatised. As mine workings became more extensive, it became increasingly difficult for workers to monitor their colleagues' behaviour, and therefore such rules were intended to foster self-regulation in the interest of collective safety. Organised labour demanded a disciplined workforce.[19]

Aside from providing their own support for sick or impaired members, unions pressed for better provision for the injured from employers. As we have seen, Durham and Northumberland miners injured at work were customarily paid allowances of 'smart money' by their employers during periods of incapacity. Workers and mine owners frequently disagreed on the basis and extent of such incapacity benefits and disputes over smart money were common. Miners preferred to see sick pay as their right, regarding it as part of their remuneration package for the dangerous work they did. Mine owners, in contrast, tended to view smart money as a 'gratuity', a generous gesture of goodwill towards employees that could be withdrawn whenever they chose.[20] Such differences of opinion were usually expressed most dramatically and vociferously during strikes. As early as 1793, striking mineworkers at Hartley Colliery, Northumberland, included in their list of demands a call for five shillings a week for injured pitmen. The strikers claimed this would bring the pit into line with the provisions made for injured miners at other collieries in the region.[21]

Despite such early efforts, north-east England miners' entitlement to smart money remained very uncertain well into the next century.[22] In 1825, pitmen from Thomas Hepburn's newly founded United Association of Colliers complained 'of having no "smart money" or weekly allowance, made to us, when we have been lamed in following the work of our employer'.[23] Even at pits where owners routinely paid such allowances, there were frequent disagreements about the amount of smart money on offer. During the great strike of 1844, disgruntled colliers in Northumberland and Durham, for instance, included a doubling of smart money from five to ten shillings a week among the list of demands they presented to mine owners.[24]

While financial mitigation of the effects of injury was a priority for many pitmen, the improvement of mine safety also played a significant role in the industrial politics of mining in the nineteenth century. Indeed, the two issues were often intimately linked in the minds of miners. When striking colliers in north-east England demanded more smart money in 1844, they publicly reasoned this would 'cause greater care to be taken of [their] lives and limbs'. For these pitmen, sick pay was not simply a matter of financial support during incapacity; it was also seen as a spur to mine owners to improve colliery safety. Miners hoped that by increasing the financial burden accidents placed on owners, an increase in smart money allowances would incentivise mining companies to invest in accident prevention measures, such as better ventilation.[25]

Indeed, worries about poor ventilation were a frequent theme in mining union complaints about working conditions at British collieries throughout the nineteenth century. When Hepburn's union stated the 'various grievances of the pitmen of the Tyne and Wear' in 1825, it not only called for smart money; it also highlighted the dangers of bad ventilation to the lives and health of miners.[26] As English colliers commonly complained of the 'pernicious' health effects of work in poorly ventilated mines, so too did colliers in other parts of Britain.²⁷ According to Mines Inspector Tremenheere, '[t]he injurious effect of badly ventilated pits upon the health of the people working in them was one of the prominent subjects of complaint put forward by the delegates of the Miners' Union in Scotland in 1844'.²⁸ Miners' demands for improved health and safety measures were especially loud following major mining disasters. Within a few months of the Hartley Colliery catastrophe in which more than 200 people died, mostly from suffocation after being trapped underground due to the mine's single shaft being blocked by fallen machinery, a petition signed by 17,000 miners was presented to Parliament urging better government regulation of mining safety.²⁹

Such campaigns were an increasingly significant and frequently successful component of organised labour's struggle for better colliery safety from the 1840s. Coalminers were among the first Victorian workers to demand state intervention in their industry.³⁰ After the enactment of the Mines and Collieries Act of 1842, miners' leaders came to realise that government could, if handled carefully, be a useful ally in industrial politics and an effective protector of workers' health and safety. This realisation was particularly influential in the wake of the crushing defeat of pitmen in the great strike of 1844, when miners were forced to rethink their strategy for bringing about change. Following the strike, miners began to turn more to the state for help, petitioning lawmakers directly rather than simply battling coal owners and their entrenched interests in the coalfields. The passage of the 1850 Mines Inspection Act owed much to this new approach, as pitmen – through the Miners' Association of Great Britain and Ireland – lobbied Parliament, assisted by allies such as radical MP Thomas Slingsby Duncombe, to great effect.³¹

As their petition after the Hartley disaster of 1862 indicates, miners commonly pointed to shocking colliery accidents in the sector to put pressure on legislators to do something to protect them.³² These had the power to provoke public sympathy and were widely reported in the press. Yet alongside disturbing tales of underground deaths, unions' growing interest in influencing industrial politics through appeals to the state meant images of industrial disease and deformity highlighted by government investigations of mine work were also powerful rhetorical tools. At the start of the great strike of 1844, speakers at a mass meeting held near Gateshead used imagery reminiscent of the 1842 Children's Employment Commission Report when they highlighted the ways in which miners' bodies had been worn down by 'toil' or hurried into 'premature old age' to support the justice of their cause.³³ Such imagery was taken up by critics of industrial capitalism more generally. In his *Condition of the Working Class in England* of the same year, Friedrich Engels actually cited the work of the Children's Employment Commission directly to document the '[d]istortions', 'deformities' and 'malformations' suffered by mineworkers because of their work in the coal industry. In his view, the damaged bodies of miners were visual evidence of the 'shameless oppression of the "coal kings"' and proof of the evils of the emerging industrial system.[34]

As Engels' and striking miners' rhetorical use of the figure of the disabled miner in 1844 suggests, such imagery was also used to garner support for causes beyond health and safety and provision for disabled workers. By the early 1860s, aided by pro-labour newspapers, such as John Towers' *British Miner* (which made a point of documenting every fatal mining accident), it became commonplace for miners and others to list 'lack of protection for our lives and limbs' alongside other grievances ranging from the truck system to disputes over the weighing of coals.[35] Many correspondents to these newspapers argued that trade union combination was the only defence against the 'appalling amount of physical suffering' experienced by men who were 'slaughtered at the rate of 1000 a year [and] endure the pauperism resulting from 10,000 permanent disablements' caused by mine work.[36] In an address to the miners of Staffordshire in May 1863, James Moon similarly cited examples of loss of life and 'fractured limbs and mutilated bodies' that left families without breadwinners and dependent on 'the pittance of the benevolent' or the Poor Law. He called for his fellow miners to 'come forward like men, and join in the association for the better protection of the health and lives of the working miners of Staffordshire'. The miners, Moon argued, were 'freeborn sons of Britain' and should not allow themselves to be treated like the 'serfs of Russia'. For Moon, union membership was a means of restoring manhood against miserable dependency.[37] In spite of evidence that many injured miners found work in the industry, supporters of combination frequently represented disability through the lens of 'suffering', as something that would lead inevitably to impoverishment and devaluation. Unionisation campaigns commonly played on fear of impairment to drum up support. In this context, as Rose has argued, the rhetoric of workplace disability served as a 'narrative prosthesis' – a crutch to support calls for political mobilisation on a range of occupational grievances.[38] While these representations employed death, disease and disability to critique rapacious industrial capitalism, they had little to say about the day-to-day lives of disabled coalminers.

As important as unions were in harnessing and shaping nineteenth-century mineworkers' fears and responses to industrial disability, they are not the whole story. The presence and impact of unions in the coalfields at this time ebbed and flowed and were never uniform. In the first half of the century, pits in the north-east of England tended to be the hotbeds of mining unionism while miners in Scotland and especially south Wales were generally slower to organise. In 1844, two-thirds of the 32,000 members of the Miners' Association came from north-east England, while most of the remainder came from Lancashire and Yorkshire.[39] As late as the mid-1860s, only a third of workers in British coalmining were unionised, and most of these were hewers.[40] By 1900, despite an impressive upsurge in membership, union density in the industry was still only around three-fifths (59.5 per cent) of the total workforce.[41] Despite their prominence in the historiography of coalmining, then, unions were not always the conduit through which industrial politics played out in the sector.[42]

Many miners acted independently on matters of health and safety, making decisions on their own or with their immediate work group colleagues about how best to protect their lives and bodies. At times, this meant they were prepared to defy colliery management and even the law. In October 1867 the *Colliery Guardian* reported the case of John Macmillan, a Durham mine-worker prosecuted for breach of contract. Macmillan left his place of work at Washington Colliery without the consent of his employers. When asked to explain his actions in court, Macmillan 'complained that the atmosphere of the pit was so hot that his eyesight and head had been greatly affected, the giddiness resulting disabling him from performing his work'. When asked if he was willing to return to work, Macmillan stood by his original decision, insinuating that 'he had a wife and four children to consider'. Macmillan, then, seems to have made his unilateral decision to leave his station out of fear for his health and physical ability to support his family. The price he paid, moreover, suggests his worries were genuine and deep-rooted. For his intransigence, Macmillan was 'sent to prison for one month with hard labour'.[43] Macmillan was certainly not alone in this regard. Edward Rymer, as we saw in Chapter 1, was also imprisoned for breaking his bond of employment after leaving a pit he saw as injurious to his health.

Unionised miners who fell foul of the law or tried to use it to mitigate the financial effects of impairment could not always rely on the support of their unions. Despite their rhetorical concern about disabled workers, labour leaders did not view *all* injured miners with sympathy. Giving evidence before a parliamentary committee on employers' liability for workplace injuries in April 1877, Benjamin Pickard, the secretary of the West Yorkshire Miners' Association, referred to the case of a mineworker burnt while working underground. The injured man had sought union assistance to sue his employer for compensation, but was refused because he had entered his place without clearance from his manager, as required by mine rules. In the eyes of Pickard and the union, this made the injured miner responsible for his own injuries, not his employer.[44] Reactions like this again illustrate how notions of who was to blame for industrial disability could have a profound effect on the experiences of injured workers. As in the rules they wrote for union accident funds, labour leaders often displayed a moralistic attitude in their day-to-day dealings with injured miners. Mineworkers deemed culpable for their own injuries because of reckless behaviour were cast off. Rather than being seen as victims of industrial capitalism whose disabilities called for solidarity and collective action, they were viewed instead as individuals whose behaviour acted as a warning to others. Like the disabled nineteenth-century American railroad workers studied by John Williams-Searle, such men 'tested [the] boundaries of brotherhood' among miners.[45] Unions, then, were not always useful, supportive or even sympathetic allies in the industrial politics of disability. Sometimes they were as recalcitrant and obstructive as the most difficult and uncaring employers.

Industrial conflict as a cause of disability

As the examples of labour unrest referred to above suggest, the British coal industry has a long history of fractious industrial relations. During the nineteenth century, most decades were marked by at least one serious and protracted dispute between mineworkers and employers. Strikes were common throughout the coalfields, major ones occurring, for example, at pits in north-east England, Scotland and south Wales in 1810, 1831–32, 1844, 1856 and 1871.[46] Such disputes left their mark on the disability histories of British coalmining communities. On a somatic level, they often helped create impairment. As Rose notes, industrial disputes through the ages have frequently turned violent, leaving many workers seriously injured.[47] The same is true for the British coal industry in the eighteenth and nineteenth centuries.[48] Accounts of strikes and protests from the period are littered with references to violent and injurious incidents. In March 1789, Thomas Barnes, an 'agent for Walker Colliery' in Northumberland was set upon by protesting pitmen who 'beat him' so badly that he was 'disable[d] ... from working for some time'.[49] Managers were not the only ones hurt in violent incidents like this. Workmen were also injured in heated industrial disputes such as the one at Ayrshire pits in 1842. Striking colliers there, according to a report in *The Times*, took umbrage to imported labour brought in to keep the mines open. Among those caught up in the tumult, were 'two men who had lately come to reside' in the area, presumably to work in a local colliery. One afternoon, on their way home after work, the two men were attacked 'by an incensed mob, who threw stones at them and seriously injured their persons'.[50] Inhabitants of mining districts from all social classes frequently fell victim to similar acts of violence.[51]

Mineworkers with pre-existing physical impairments were not automatically insulated from workplace violence. In February 1832, collier David Edwards went to work in a coal pit attached to an ironworks at Blaina, south Wales. Many workers there were engaged in a dispute with their employer and were effectively on strike. Regarded as a blackleg, Edwards soon came to the attention of the notorious 'Scotch Cattle' – groups of Welsh miners who terrorised fellow workers that refused to take part in collective action – and received a beating for his troubles. When he gave evidence about his experiences in front of a local magistrate, Edwards declared that around the time of the attack he was 'lame and not able to walk as fast' as other workmen. Although having a mobility impairment did not ultimately save Edwards from violence, it may have stopped the attack from becoming more serious, as during the assault one member of the gang urged restraint on the others, saying Edwards 'was not guilty to be punished'. The court records remain unclear on this point, however, and it may be that the Cattle's leniency towards Edwards was more to do with fact that he had decided to stop work before the attack than his 'disability'.[52] Whatever the case, it is clear that injured miners like Edwards could be hurt in industrial unrest just as other workers were. The passions aroused during heated disputes meant no one was completely immune from physical attack, not even people who in other contexts might expect to be treated with sympathy and kindness.

Disabled mineworkers were not only the victims of violence; sometimes they were involved in meting it out to others – and to great effect. In his memoir of his life as a mineworker, self-declared 'half-blind' and 'cripple' Edward Rymer recalled a physical altercation he had with one of his supervisors during a pay dispute in 1859. According to Rymer, his superior became aggressive and 'tried to force me out of the pit by an assault'. Rymer did not submit passively but 'retaliated' and a 'scuffle' ensued in which his surprised manager came off worse (in Rymer's reckoning, at least).[53] Incidents such as these are a powerful reminder of the agency of disabled workers in industrialising Britain. Despite popular accounts of the Industrial Revolution, in which disabled people tend to feature as the passive victims of disruptive economic forces, this example suggests that not all workers with impairments were willing to be, quite literally, pushed from the workplace without a fight.

Some nineteenth-century Britons thought industrial disputes weakened the bodies of mineworkers in less direct ways. A witness to the great pitmen's strike of 1844 in north-east England claimed that many strikers 'so much reduced themselves by low living, that they were good for nothing for weeks after they started [work] again'. Inspector Tremenheere concurred with this view when he also reflected on the effects of the strike on the bodies of miners, noting that 'it was a long time before they recovered their former strength'.[54] Nearly three decades later, the secretary of a Welsh coal owners' association, Alexander Dalziel, made similar remarks in his observations on the south Wales' miners' strike of 1871. The 'diminution of the physical power of the workmen' caused by the dispute, Dalziel wrote, was so serious that it took two months after their return to work 'before they could be said to have been restored to good working condition'.[55] Late Victorian newspaper accounts of strikes sometimes depicted miners and their families as being so weak with hunger that they could not walk.[56] Moreover, in an industry where the 'seasoning' of workers' bodies was important to making them strong and flexible enough to cope with the rigours of underground labour, absence from work, whether voluntary or enforced, could result in reduced somatic capabilities. In the view of many mine owners, managers and inspectors, the weakened physical capacities of mineworkers brought on by their idleness and abstemious living during strikes meant that prolonged industrial action was a cause of 'disability' (albeit temporarily) in its own right.

Strikes were also thought to increase the occupational risks to which miners were exposed on their return to work. In April 1827, British newspapers reported the deaths of at least five Scottish colliers when they inhaled noxious 'choke damp' following the reopening of a pit after a recent strike. The reports implied that the fatalities were a tragic consequence of the dispute because the cessation of mining operations during the strike had allowed deadly gases to build up through a lack of ventilation.[57] The importation of blackleg labour to break strikes also aroused the suspicion of former strikers, who claimed it compromised their safety. In the aftermath of the Haswell Colliery disaster of September 1844, old hands employed before the strike of that year complained that the 'strange men' brought in from outside local pit villages to end the dispute might be to blame for the calamity. Their inexperience of conditions in the coalfield, so the argument went, meant they were incapable of spotting the tell-tale signs of underground dangers. This made them a liability not only to themselves, but also to the safety of other mineworkers.[58] Such claims were common, though it is difficult to gauge their veracity given the acrimonious post-strike contexts in which they were often made. Yet, as Robert Colls points out, the 'disastrous aftermath' of the strikes in north-east England of 1831–32 and 1844 'offers some support for the "true-bred" pitmen's claims'. In the six months from May to November 1832, for example, four 'major explosions' occurred in the region.[59]

Disabled people and industrial conflict

Industrial disputes may have caused injuries, but what happened to people who already had impairments during such tumultuous times? The experiences of disabled people in coalfield communities during periods of industrial unrest were mixed. In some cases, intense conflict may have presented disabled mineworkers with better work opportunities; in others it resulted in increased hardships. For the courageous or desperate miner willing to work as a hated blackleg, strike-breaking could be a profitable, if risky, business. Accounts of nineteenth-century miners' strikes make clear that many mineworkers were indeed prepared to take such risks. Blackleg labour helped break numerous strikes, much to the chagrin of unionised workers.[60]

Miners with impairments were among those who took advantage of the opportunities for work afforded by strikes. Recall, for instance, the example of 'lame' collier David Edwards cited previously who went to work in the 1830s at a mine in Blaina despite workers there being on strike. Judging from his testimony about his encounter with the 'Scotch Cattle', it seems Edwards had been looking for work in the area unsuccessfully for some time before he took up employment at the pit.[61] If Edwards' mobility impairment undermined his job prospects at collieries with relatively harmonious industrial relations, in the context of the dispute at Blaina it seems to have presented no major barrier to his participation in the mining workforce. Edwards was certainly not alone; other 'disabled' mineworkers also crossed picket lines to take up work during strikes. In his account of the 1844 strike in Northumberland and Durham through which he had lived, historian Richard Fynes recorded a fatal accident involving a visually impaired Welsh collier known as 'Blind Davy', who died after falling down a mineshaft. As one of the blacklegs brought in to break the strike, Davy's death was apparently little mourned by fellow workers, with one 'old furnaceman' remarking that the dead man was 'only a Welshman ... we[1]l out of the way'.[62] Despite the opprobrium blacklegs attracted, for many miners with impairments, like those without, the opportunity to make money as strike-breakers was just too enticing to shun. Moreover, although hated by striking miners, men like Edwards and 'Blind Davy' were probably quite attractive to employers determined to defeat unionism. Both men may have had impairments, but they also seem to have been workers with prior experience of mine work. During a strike, such experience was often at a premium as labourers with any kind of familiarity with mining could be very hard to find. Recalling the strike of 1844 in north-east England, coal viewer Ralph Elliott reported that the inexperienced Irish labourers imported to break the strike 'were so awkward at the work' that 'they could scarcely earn what their food cost us'.[63] The economics of strikes, then, may have improved the position of disabled miners in the labour market and allowed them to command better wages than they could in less turbulent circumstances.

Strikes were not the only form of industrial action practised by mineworkers. Other less extreme and disruptive methods were used to influence the terms of work at nineteenth-century collieries. A common tactic employed by colliers, especially in Scotland and north-east England, was the restriction of output. By limiting production, mining unions aimed to manipulate the price of coal, and hence miners' wages, while simultaneously maximising the number of their members in work.[64] As Victorian commentators remarked, this policy had a levelling effect in that it allowed less productive miners to earn the same as the most productive. This is because it forced miners capable of producing more than the set limit to work below full capacity, while ensuring good wages for those who struggled to reach higher production quotas (due to the inflated piece-work rates the policy encouraged). Writing in the 1840s, Mines Inspector Tremenheere outlined the policy of restriction then practiced by Scottish colliers and commented on its effects. It was based, he claimed, 'on the irrational principle of allowing no one man to do more work than another; of forcing an unnatural equality of earnings on the young and the old, the strong and the weak, the industrious and the idle'.[65] Despite Tremenheere's disapproving tone, his observation clearly suggests the potential benefits of restriction for 'disabled' miners. Where restriction of output was in force, noted an official from Clyde Iron Works in 1845, older 'failing men' who normally relied on the assistance of their sons and others could produce as much coal as others of greater strength. A decade later, a 'gentleman of long experience' in the Scottish coalfield similarly affirmed that where restriction was practiced 'infirm old men' were able to earn the same pay as much stronger young colliers.[66]

Restriction was favourable for disabled miners in other ways too. As Colls has noted, union limits on output also facilitated 'the "carrying" of men who needed help'.[67] In other words, because some miners worked below full capacity, there were workers underground with spare time and energy on their hands to help less able colleagues. Restriction, then, mirrored the 'protective mutuality' practised by other workers in industrialising Britain.[68] Recognising that restriction potentially helped miners with impairments make a living from mining also illuminates how industrial politics could shape the 'somatic flexibility' available to them discussed in Chapter 1. For many miners, their ability to work in mining was often just as much about the political conditions at their pits as it was about the state of the economy or their physical capacity for hard labour.

On the whole, despite the advantages some disabled mineworkers enjoyed during periods of industrial conflict, protracted disputes between mine owners and their employees tended to cause severe hardships for mining families. Such hardships often affected disabled people particularly badly. In matters of industrial politics, courts commonly upheld the interests of mine owners and their representatives at the expense of workers. As we have seen, judges frequently imprisoned miners for defying employers and Edward Rymer's experiences indicate that men with impairments also shared this fate. Yet the courts were not the only state institutions used to discipline insubordinate or troublesome mineworkers. Sometimes the Poor Law was also utilised as a weapon in industrial disputes, and poor relief withheld from striking miners.[69] While local welfare officials had a legal duty to relieve destitution, the Vagrancy Act of 1824 allowed for the punishment or denial of relief for those who refused work they were capable of performing. This made it legally permissible for Poor Law officers to deny public welfare to strikers or workers who refused to participate in strike-breaking. For example, in 1875 the Merthyr Tydfil Board of Guardians turned down an application for relief from a number of colliers from the Plymouth and Cyfarthfa works because the applicants would not accept work at a nearby colliery at Dowlais. The reason for the men's refusal was that work there had already been 'refused by the Dowlais Colliers', who were in dispute with their employer, and they did not want to undermine the cause of their fellow miners.[70] Mineworker solidarity was clearly something Merthyr Tydfil welfare officers did not want to be seen supporting.

Examples like this reveal the politicised nature of welfare provision in Britain's coalfields. As Jamie Bronstein has observed, industrial paternalism in nineteenth-century Britain was based on the expectation that workers in need of medical care or financial support showed deference and gratitude to their benefactors and social betters.[71] By resisting the will of their employers, strikers were the antithesis of the 'good' workers worthy of kindness and compassion mine owners sought to cultivate. Consequently, miners in dispute with their employers were more likely to be regarded as enemies to be crushed than people who 'deserved' help. As the preceding chapters have shown, it is inaccurate to automatically equate disability with dependency. Nevertheless, physical impairment did cause financial difficulties for many mining families and these could be ruthlessly exploited by mine owners who sought to defeat unionism.

Determined mine owners frequently stopped sickness benefits, or smart money payments to injured workers when they were in dispute with their employees. Such was the case with Thomas Lawton, a seventeen-year-old who had been badly burnt in a mine explosion in south Durham. Because of his injuries, Lawton had been unable to work for half a year and had received smart money during his period of incapacity. When his pit community got caught up in the industrial turmoil sweeping the region in 1831, however, his previously benevolent employer had a change of heart and stopped payments. Although Lawton had been off work and could not, therefore, practically go on strike, the fact that his fellow workers chose to do so was enough for his employer.[72]

As this case shows, mine owners were often prepared to use any means available to destroy the morale of mining communities during strikes. The vindictiveness some owners and their managers showed towards those with the temerity to challenge them frequently impacted on disabled people. In 1853, Edward Richardson was one of several men at Seaton Delaval Colliery who lost their jobs for striking. Perhaps wishing to make a special example of Richardson, who had been a union agitator during the 1844 strike, the colliery agent chose to punish the dismissed worker's family further by withdrawing the smart money paid to Richardson's son, Matthew, a hewer at the colliery who had been injured in an accident some months before. When Richardson went with Matthew, 'who was then walking on crutches' to ask 'what he intended to do with his son who was disabled from work', the agent replied 'we have nothing to do with you or your son, you have brought it all on yourself, you might have been in a better situation than you are, had you looked towards your own interests'.[73] In north-east England, owners continued to use smart money as a weapon in industrial conflict like this into the 1880s and beyond.[74]

While striking mineworkers did not use disabled people as pawns in their industrial disputes in quite the same way as some employers did, their actions could undermine the well-being of injured colleagues on occasions. In the heat of battle with intransigent and ruthless mine owners, labour leaders were often willing to sacrifice the welfare of disabled people to achieve victory. It was not uncommon, for instance, for cash-strapped unions to suspend or reduce sickness benefits to facilitate the continued support of striking miners.[75] Some friendly societies might similarly come under financial pressure during industrial unrest. In the first half of the nineteenth century, the distinction between trade unions and friendly societies was not always clear-cut. During industrial disputes mine owners and their supporters, like the pseudonymous writer 'Scrutator', commonly complained that friendly societies were really unions in disguise, used to promote collective action. In a pamphlet published in 1832, Scrutator reflected on the miners' strike of the previous year and argued that a friendly society supposedly set up for the 'relief of the sick or the support of the aged' had actually been used to fund strikers in what he contemptuously termed their 'unmanly slothfulness'.[76]

Despite the partisan nature of his comments, Scrutator had a point. Striking miners *did* sometimes utilise friendly society funds to support themselves, especially during lengthy disputes such as the strikes in north-east England of 1831 and 1844. According to a report by the Inspector of Mines, after exhausting their own resources, including selling or pawning personal belongings, strikers in 1844 were reduced to such a desperate state that they broke up their 'benefit clubs' and used the funds for their immediate support. This obviously had a very serious impact on the sick and injured miners who relied on club benefits to maintain themselves and their families. As the inspector realised, moreover, this was not a short-term problem. It would take years before the friendly societies involved replenished their funds to pre-strike levels, if ever.[77] As prospective members with pre-existing medical conditions were usually barred from joining friendly societies, this situation was a major blow to many people with impairments in the coalfield who had depended on assistance from these clubs. Unless the societies resumed business and accumulated sufficient funds again, they were unlikely to ever receive cover under similar schemes in the future.

Friendly societies were also used by mine owners as a vehicle for achieving supremacy in industrial politics. Govan Colliery Friendly and Free Labour Society, instituted in 1826 on the insistence of the colliery's proprietor, ironmaster William Dixon, was regularly held up as a model for promoting an obedient mining workforce. Dixon had recently defeated a strike by mineworkers and was determined to eliminate any prospect of future disruptions at his colliery. After 1826, all workmen employed at his pit were required to join the new society. As its name made clear, the society was for the benefit of *free* (as opposed to unionised) labour only. In line with other mine owner-encouraged pit clubs of the time, such as Hetton Colliery Agents and Workmen's Friendly Society, the Govan Society expressly prohibited trade union members from joining. By barring unionists from the society while simultaneously making membership of the scheme a condition of employment, Dixon created a significant barrier to the establishment of an effective union at his colliery. For those who joined the society, moreover, any suspicion that they were even a little involved in union activism risked jeopardising their access to sickness and injury benefits in the event of incapacity. In return for their loyalty to their employer, members were offered security against strikes as well as sickness, with the society guaranteeing payment of benefits to members unable to work due to intimidation.[78]

During his battle with striking miners in 1825–26, out of which his plans for a Free Labour Society grew, Dixon employed a dual strategy to achieve victory. As other British colliery owners did repeatedly throughout the nineteenth century, Dixon employed blackleg labour to break the strike and evicted striking miners from company-owned housing. Evictions freed up accommodation to house imported strike breakers and aimed to undermine the resolve of employees who refused to return to work.[79] This tactic was particularly used by employers during miners' strikes in Scotland and north-east England, where colliery-provided housing was more prevalent than in south Wales.[80] Many mine owners were prepared to do whatever it took to forge a docile and obedient labour force – even if that meant throwing vulnerable people out of their homes. Recounting employer tactics during the strike of 1844 in north-east England, Fynes wrote that '[w]holesale turning to the door commenced in almost every colliery village; pregnant women, bedridden men, and even innocent children in the cradle were ruthlessly and remorselessly turned out.' In his view, the eviction of 'the aged, the sick, and the feeble women from the homes of their childhood' was part of a deliberate strategy by employers to provoke striking miners to 'break the peace' so that the full force of the law could be brought down upon them. Among those evicted at Pelton Fell were an 'old blind woman, 88 years of age, who was left exposed to the cold and rain', while at another colliery 'a young man to whom a misfortune had happened was ruthlessly put to the doors'.[81]

Families turned out of their colliery-owned accommodation sometimes found shelter in the homes of friends, relatives or charitable neighbours. Such sources of assistance were inadequate to meet the needs of everyone evicted during particularly bitter and protracted disputes. During the dispute described by Fynes, many striking miners and their families were forced to camp by roadsides in makeshift shelters like the one pictured in the *Illustrated London News* in August 1844 (Figure 5).[82] The image of homeless strikers shivering in encampments became one of the most powerful symbols of the 1844 strike. 'It is true that the masters had a right to do what they liked with their own,' wrote Fynes, 'but on the score of humanity and fellow-feeling they might have refrained from turning their old servants to the doors till they had new ones ready to occupy their places.' Throughout Durham and Northumberland 'there were thousands of cottages tenantless, whilst their late inmates were camping in the open air, exposed to the inclemency of the weather'. It was, concluded Fynes, 'a cruel and dastardly revenge'.[83]

The depiction in the *Illustrated London News* of a man sitting outside a makeshift shelter holding a crutch, used the figure of the disabled miner to stand for the sufferings of the evicted in general. For supporters of the strike such as Fynes, eviction of the 'weak', pregnant, sick and old broke a moral

Figure 5 'Pitmen Encamped', *Illustrated London News*, 3 August 1844. Copyright Illustrated London News Ltd/Mary Evans.

contract that held that masters should behave with compassion towards the vulnerable in pit communities. Since employers had often demonstrated their paternalist care for their workers by allowing widows and injured mineworkers to stay in company housing after they had been injured, the depiction of a disabled miner turned out during the strike shows how easily such benevolence might be withdrawn. It was a tactic used again in later disputes. For example, in November 1865, an anonymous writer and possible participant in a long-running strike at Cramlington Colliery in Northumberland informed the Home Secretary that the families of striking miners, including those containing 'the blind, the lamed, and the sick', had been forced from their homes. Among those evicted, for example, was 'one poor creature who was in the habit of taking fits'.[84]

Highlighting the sufferings of disabled people during industrial conflicts was undoubtedly a powerful means of criticising the cruelty of employers and their supporters. In Fynes' opinion, such behaviour was 'unmanly', went against the principles of Christian charity and was unpatriotic conduct from men 'with British hearts beating in their bosoms'.[85] Published in 1873, Fynes' account of the 1844 strikes echoed coverage of strikes in late nineteenth-century newspapers which used sensationalist images of suffering, particularly hungry wives and children, to evoke sympathy for the victims of industrial conflict. Stories of the 'vulnerable' members of coalfield communities receiving equal 'punishment' to the men on strike, were clearly designed to have an emotional impact.[86]

Nevertheless, although there is evidence that disabled mineworkers may have fared especially badly during industrial disputes, such shared hardships may have allowed disabled miners to regard their 'sufferings' in a positive light. Although some may not have actually been working at the time strikes commenced, the extra difficulties they endured because of poor industrial relations may have allowed disabled miners to feel solidarity with their mining colleagues and part of the wider struggle in which their communities were engaged. The attitude of striking miners' wives is suggestive in this regard. Although they themselves were not on strike, strikers' wives were active participants in coalfield industrial disputes. In the early 1840s, an eyewitness to a strike of Staffordshire miners noted the great 'distress' they experienced and marvelled at the steadfastness of their wives in the face of such difficulties. 'The women, particularly,' he observed, 'were exceedingly inveterate in urging their husbands to hold out, saying, that they would rather live on potatoes and salt than give in.' Sacrifice and suffering enabled these women to show their commitment to the cause and membership of their working-class community. There is no reason to assume that the same did not hold true for disabled people caught up in the communal hardships associated with industrial action.[87] Indeed, the ruthless and deliberate manner in which some mine owners targeted disabled people suggests employers certainly regarded them as aligned with defiant miners and therefore worthy of punishment. By sanctioning the breakup of benefit clubs during strikes, labour leaders similarly seem to have viewed sick and injured miners as participants in class struggle who were also expected to make personal sacrifices for the greater good. Whether or not they actively sought it, then, many disabled people in mining districts were undoubtedly united with mineworkers in a solidarity of suffering during industrial disputes that may have helped foster intense feelings of pride and belonging.

Taking action: compensation and liability

Sharing in the suffering of evicted strikers and providing moral support for those taking part in industrial action were indirect means by which those disabled from working could demonstrate solidarity with their non-disabled neighbours. By the 1870s a more direct form of action on the part of injured workers and their families – the right to seek compensation in cases where employer negligence may have caused disablement – became a topic of increasing debate. This culminated in the passage of the 1880 Employers' Liability Act, which clarified the legal right to redress on the part of those who might suffer injury or disablement at work on account of their employers' negligence.

In the eighteenth century the families of those who lost their lives in workplace accidents could bring a lawsuit for forfeiture of property if it could be proved that the employer was negligent. The law was amended in 1846 to give dependents a financial claim in cases of accidental death where employer negligence could be proved. This was reinforced by the 1850 Mines Inspection Act, which (following provisions made in the 1844 Factory Act) opened up the possibility that fines imposed on coal owners under the law might be used to compensate the relatives of workers killed in fatal mining accidents.[88] For the most part, in spite of the large number of fatalities in mining, litigation proved too expensive for most miners to pursue against their much more powerful employers.[89] Furthermore, the doctrine of common employment that stipulated that no legal action could be brought if an injury or fatality was caused by a co-worker (a 'fellow servant') to the injured party, made actions difficult to bring. The definition of a 'fellow servant' was so loose that it left open the possibility that defence lawyers might plausibly argue that managers were also in common employment with manual workers therefore undermining any claim for compensation workers and their families might bring.[90]

Although the 1850 Mines Inspection Act did not result in a large number of compensation claims, the law may have encouraged some miners and their representatives to push for similar legal rights to seek financial redress for workers injured at work as for those killed on the job. In 1854, for example, as the 1850 Act came up for renewal, the Northumberland and Durham Miners' Association petitioned the Home Secretary to empower mines inspectors 'to institute proceedings at Law to recover damages' from mine owners for the *death or injury* to any miner ... caused by the neglect of the owners or their agents'. In such cases, the association urged, any awarded damages ought to go 'to the party injured'.[91] Given the costs of litigation (estimated at a minimum of £30), the association proposed that the state should shoulder the financial burden, citing the fact that factory inspectors already had the power to bring legal actions on behalf of injured workers and their families.[92] While the miners' calls were rejected on this occasion, further demands for more effective compensation legislation resurfaced in the aftermath of the shocking Hartley Colliery disaster of 1862.[93]

Despite the considerable barriers, some injured mineworkers and their supporters did seek compensation from their employers through the courts. Bronstein, for example, cites the mid-nineteenth-century case of a Scottish collier boy who was awarded damages after his leg was amputated following an accident caused by his employer's neglect of safety.[94] Successes like this, however, were quite rare. More typical of the experiences of disabled mineworkers before the courts was that of Ayrshire miner Thomas Mclachlan. In 1855 Mclachlan suffered a brain injury after a 'fire lamp' grate fell on him as he descended the pit. This had 'rendered [him] incapable of supporting himself and his family'. Being 'obliged to go on the parish', Mclachlan sued the company that had employed him for £1000. The issue of liability hinged on whether the grate had been adequately secured, but in spite of the allegation that one of its four legs was 'either broken off or loose', the jury returned a verdict for the coal company after just 'a few minutes deliberation'. The press report of this case referred to it as 'another coal-pit accident action', suggesting that, although their chances of success might be slim, other disabled miners around this time similarly had the courage to take on their employers in court.[95] This was as true for other parts of Britain as it was for Scotland.

In 1858 a mine sinker in north-west England named Griffiths won £100 in compensation at the Liverpool Assizes after he was badly injured when a bucket fell down a mineshaft and hit him while he was working. The case, which again hinged on the argument that the 'machinery was imperfect and insufficient', received lengthy attention in the *Colliery Guardian*. This was sympathetic to mine owners and its coverage of the judgement suggested that

Griffiths' victory posed a significant challenge to their usually successful efforts to avoid liability for workplace accidents. Believing the case set an important legal precedent, the *Colliery Guardian* urged its readers to learn lessons from the lawsuit. Employers ought to take the utmost care in ensuring that safety equipment was used and in good working order at their pits and that 'all proper regulations' were enforced. If they did not, there was a great danger that litigious miners and their unions might abuse the law 'just as the law for the protection of railway passengers is often perverted into the means of extortion' by people injured in railway accidents. The journal also restated the commonly held view that workmen were paid for the risk-taking their jobs entailed as well as the labour they performed. Given this, it implied, the relatively high wages mineworkers received were compensation enough for any injuries they might sustain at work. Furthermore, as 'the great majority of colliery owners aid in providing their workmen with medical attendance and the means of subsistence when hurt under their employ', why would injured miners need to seek damages – they were already adequately provided for by their employers. Or so the *Colliery Guardian* thought.

Mineworkers like Griffiths clearly had a different opinion and in the second half of the nineteenth century increasing numbers of those hurt in mining accidents or their representatives began to seek legal redress – first in the courts and then through Parliament. When it reported the Griffiths case in April 1858, the *Colliery Guardian* observed that in 'some districts working colliers are amazingly fond of law'. It was due to this fondness for legal action that the *Colliery Guardian* reasoned Griffiths' example would embolden restive mineworkers to seek damages from their employers whenever someone was injured in a colliery accident.[96] Such fears were ultimately unfounded as Griffiths' challenge to mine owners' legal supremacy proved short lived. Within a few months of his victory, Griffiths' employer's legal team managed to get the initial ruling overturned at a retrial by successfully invoking the fellow-servant rule to argue that the colliery was not liable for his accident.[97] Griffiths' eventual defeat shows just how determined mine owners were to defend their advantageous position under the law and the superior resources they had at their disposal to do so. Against such opponents, miners faced an uphill struggle to secure recompense for injury. Yet, fight they did.

As Bartrip and Burman have argued, by the 1870s it was becoming evident that state regulation of coal mines and other dangerous workplaces could only go so far in reducing the risk of death or disablement. Other methods were also needed to improve safety. Echoing the argument of English pitmen who called for substantial increases in smart money during the great strike of 1844, lawmakers came to see the threat of compensation as a way of incentivising employers to take better care of their workers' health and safety.[98] This change in attitude was reflected in the provisions of the Coal Mines Act of 1872. This consolidated mining regulations into a single law and indicates the trajectory of health and safety legislation relating to the sector after 1850. By the early 1870s, the principle that fines imposed on mine owners for breaching safety legislation could be used to compensate the relatives of those killed in mining accidents had expanded to include miners injured in the workplace – provided the accident was not the fault of the worker. To be sure, the 1872 Act did not give disabled mineworkers (or the relatives of those killed at work for that matter) an absolute right to compensation. Decisions regarding the use of fines for the benefit of accident victims or their families were left to the discretion of the Home Secretary.[99] But the law did signal that policymakers were becoming increasingly concerned about industrial injuries as well as fatalities. It also indicated that, by the latter part of the nineteenth century, the British state had developed a two-pronged approach towards mining accidents that incorporated prevention and compensation. Thus, the likelihood of accidents happening was to be reduced through improved safety measures and, when they did occur, those affected were to be adequately compensated.

The growing pressure for legislation to clarify workers' right to legal redress in the event of employer negligence in the 1870s owed much to the extension of the electoral franchise in 1867, which increased the electorate by around 1.5 million people. It also reflected the growing confidence and ability of trade unions to speak with a united voice following the formation of the Trades Union Congress (TUC) in 1868.[100] By 1880, mining unions had come to regard taking legal action on behalf of miners killed or hurt in accidents as part of their remit. For example, from 1870 the Durham Miners' Association ruled that in every case where a managers' negligence had caused the death of a member, the union should seek compensation through the courts.[101] The TUC discussed a compensation bill in 1874 and 1875 and, in 1876, Alexander MacDonald, the president of the Miners' National Association and Liberal MP for Stafford, introduced a bill reflecting workers' demands for compensation to Parliament.[102] Following strong opposition from mine owners, the bill was withdrawn and the question of employers' liability for accidents in the workplace referred to a select committee. But after the landslide Liberal victory in the 1880 general election, in which compensation had been an important issue, a bill was introduced that would become the Employers' Liability Act of September 1880. The passage of this legislation has received considerable attention from scholars, and it is not the place of this chapter to re-tell this story.[103] However, the debates around compensation merit some discussion as they reveal contested responses to disabling injury. As people who profited from a dangerous trade, most coal owners opposed any tightening of the law that might make them financially liable for accidents in the workplace. Yet the debate was more complex than a simple pitting of the interests of 'capital' against those of 'labour'. Many workers shared employers' fears that new legislation would cost livelihoods and undermine existing systems of support for sick and disabled miners.

Those opposing legislation defended the principle of paternalist care that had provided support for injured miners for decades that, they argued, had fostered co-operation between masters and men. At the same time, they cast suspicion on those injured or disabled miners who might try to exploit any new law. Following a deputation of the Miners' Association to the Home Secretary to support Macdonald's first attempt to introduce a bill in 1876, the Liberal MP for South Durham, Joseph Pease, warned that legislation would 'provoke a feeling of antagonism between employers and employed', for it would often be impossible to determine who was to blame. Ignoring the fact that disputes over the provision of smart money and other benefits had at times themselves caused 'antagonism between employers and employed', Pease reasoned that '[i]f the master was to be responsible for every crushed finger, or anything of that kind, the "smart money" which was a great object to the men, would be taken away, and they would lose the help which they otherwise got when they were out of work.'[104] A letter to the *Colliery Guardian* in 1878 warned that extending the law on compensation would merely amplify existing abuses of colliery sick funds as 'many of the lazy class of men try their best to abuse [the system] by endeavouring to force the visiting officer to believe that the ailment they are suffering from was produced by an accident in the pit', when on medical inspection it was proven to be the result of some 'organic disease'. Sick funds, it argued, were policed by the men themselves, who had an interest in preventing fraud as it threatened the solvency of the schemes to which they contributed out of their own wages. However, if a 'colliery proprietor' became solely liable for an injury a workman received 'all his fellow workmen would be only too willing to swear black was white in his favour'.[105]

Some politicians sympathetic to the coal trade sought to head off the Employers' Liability Bill by proposing their own solutions. These sought to avoid the potential for antagonism between employers and workers by extending into law the principle of co-operation exemplified in the permanent relief fund movement and other voluntary schemes. For example, during a parliamentary debate in June 1878 reviewing mining legislation, Newcastle MP Joseph Cowen proposed the creation of a fund similar to the one established for keelmen working on the River Tyne in the eighteenth century. Financed by a tax on coal and compulsory contributions from colliers, this fund would provide an income for disabled mineworkers unable to follow their occupations. Cowen pointed out that such a scheme would be in the public interest, as it would help reduce the poor rate. Glamorgan MP Hussey Vivian, a strong opponent of employers' liability legislation, similarly urged the Home Secretary to consider establishing a fund 'for the support of widows and children' of those killed in mining accidents. Macdonald disagreed, however, arguing that the financial consequences of mining disasters for their victims ought to be 'met by the providence of the people rather than by the funds of the State'. While he supported voluntary welfare schemes that encouraged miners to supply 'the wants of their family by their own foresight and thrift', Macdonald's opposition to using state funds to provide support for the injured and their families was based on a belief that such measures would allow employers to shirk their responsibilities towards their workforce. In his view, only a compensation bill would force employers to take their duties to ensure workers' safety seriously.[106]

Nevertheless, the potential costliness of compensation was a key issue for opponents of legislation. However much compensation was intended to reduce the costs of mining accidents to the public, some argued that it would inevitably place burdens on public finances since mine owners might face a glut of ruinous claims after large-scale disasters.[107] Beyond this, employers and miners alike feared that an increase in lawsuits would cause bad blood that would sour industrial relations and halt the co-operation that had advanced as a result of initiatives such as the permanent relief fund movement. John Bryson, leader of the Northumberland Miners Association, told the crowd at the Northumberland Miners' Picnic in 1880 that it was better for miners not to risk losing their hard-won benefits and to press for changes to safety at a local level rather than through Parliament.[108]

Speaking alongside Bryson at that gathering was former miner Thomas Burt, now a Liberal MP, who argued that schemes such as the permanent relief fund, though admirable, were merely voluntary. For Burt, compensation legislation would place financial protection for workers seriously injured in the workplace on a much surer footing.[109] Against the view of the disabled miner as fraudulent or responsible for his own injuries through carelessness stirred up by opponents of reform, trade union supporters of legislation spoke instead of the 'great sacrifice of life and injury to limb' caused by the 'industrial work of the country'. Speaking at the Durham Miners' Association's annual Gala in 1880, the association's president, William Crawford, described this situation as a 'standing disgrace to civilisation'.[110]

Just as trade unionists and other labour advocates had done in earlier decades, those in favour of an Employers' Liability Act used the familiar trope of disability as honourable sacrifice to advance their cause. They argued that, like soldiers wounded in battle, disabled miners had similarly been hurt performing a valuable and essential public service for which they ought to receive recompense as a matter of *right*. Mining was so important to national prosperity that those injured at work in the sector through no fault of their own deserved the respect, gratitude and support of the nation. Above all else, reformists wanted to use the state to ensure that employers were held accountable for the diseases and injuries for which their industries were responsible.

In theory at least, this is what the 1880 Employers' Liability Act delivered. The new law made an employer responsible when a worker was injured as a result of faulty or unsafe machinery that he had provided, or as a result of orders he or his representatives had given.[111] In practice, however, the Act's provisions were limited. The burden of proof was on victims to show negligence, and compensation payments were only allowed to last for three years after an accident. Furthermore, many miners were encouraged to opt out of the provisions of the Act or join insurance schemes, in return for coal owners' continued support of accident funds.[112] In December 1880, for example, two-thirds of the membership of the Lancashire and Cheshire Miners' Permanent Relief Society voted in favour of coming to an agreement with employers not to pursue claims under the Employers' Liability Act in return for an increase in 'disablement' pay from eight to ten shillings per week.[113] Indeed, as Bartrip and Burman have argued, many miners viewed the legislation as largely symbolic, a means of encouraging mine owners to commit to bettering safety rather than as a practical scheme to compensate disabled workers.[114] While the Employers' Liability Act of 1880 set a precedent for subsequent and more comprehensive compensation legislation in the immediate aftermath of its enactment, provisions for disabled miners still rested primarily on local, voluntary initiatives.

Conclusion: a legacy of exclusion?

Disability has rarely featured in histories of labour and industrial politics in nineteenth-century Britain. However, as this chapter has shown, the recognition that mining produced injury on a significant scale became a powerful driver of industrial regulation at a national level and affected industrial relations within the coalfields. As the voices of labour became louder during the nineteenth century, the 'sufferings' of the disabled miner were used to hold employers to account for poor working conditions, push for legislative responses to improve health and safety, and to rally the labouring classes to fight oppression. This rhetoric illustrates powerfully the place of disability in fostering working-class mutuality, reflected further in trade unions' own efforts to establish sick funds to help their members in times of need. Unions, particularly in north-east England, played an important role in holding employers to account for the payment of paternalistic benefits to disabled mineworkers such as smart money. By the 1870s, a joint committee of the Durham Coal Owners' Association and the Durham Miners' Association was meeting regularly to discuss worker grievances, including disputed payments of smart money to the sick and injured.[115] By the end of our period, unions were also helping miners and other workers seek redress in the form of compensation for accidents in the workplace. However, unions were not always friends of disabled mineworkers. Their vision of working-class mutuality was one based on disciplined co-operation, and their support for the injured depended on the causes of their impairment. Miners with injuries caused by their own folly or misconduct, or whose behaviour threatened the safety of their colleagues, found themselves ostracised. Furthermore, mineworkers with impairments who might seek to exploit opportunities for work during strikes could find themselves treated with the same vilification as other blacklegs.

There can be no doubt that the susceptibility of coalminers to accidents at work was a far more pressing political matter by the end of our period than it had been in 1780. Over the course of the period the idea that workers in dangerous occupations willingly accepted the risks to their lives and limbs shifted, first in response to the employment of women and the very young (both of whom Victorians often regarded as less capable of making 'rational' decisions about risk by dint of their gender and age), and later as part of a new ideology of industrial efficiency that saw accidents as a threat to productivity. Although mining remained very dangerous, miners' mortality rates fell in the fifty years following the establishment of the Mines Inspectorate from 16.0 per thousand in 1851 to 11.5 per thousand in 1900–02.[116] Better documentation of non-fatal as well as fatal accidents during this period was intended to force employers to confront their safety records, while attempts to tighten the law on employers' liability aimed to deter poor safety practices as much as provide compensation for the injured.

These measures were undoubtedly intended to benefit all coal workers and reduce occupational injury. Yet in the long run, these advancements in safety and support may have had consequences for the employment prospects of miners with impairments and chronic illness. In the first place, government health and safety initiatives undermined mineworkers' independence to determine their own ways of working. Policymakers realised that the creation of safe working conditions depended, to a large extent, on the regulation of labour.

By 1872 mine safety legislation placed a legal responsibility on owners to establish 'special rules' for their employees. These were supposed 'to prevent dangerous accidents and to provide for the safety and proper discipline of the persons employed in or about' their mines.[117] Managers and owners exploited legislators' injunction to ensure the 'proper discipline' of workers to its fullest. From the mid-1850s onwards, special rules were increasingly used at collieries to curtail the autonomy of mineworkers, much to their annoyance. To ensure compliance, moreover, supervision of the workforce increased.[118] Such developments, while pertaining to the safety of all mineworkers, helped to reduce 'somatic flexibility' in mining and may have made it harder for some disabled miners to continue working.

Second, the development of compensation, intended to provide financial support for injured workers, may have actually exacerbated the troubles of older and disabled miners by encouraging their exclusion from colliery work. At the end of the century, Edward Rymer predicted that compensation laws for industrial workers were 'likely to be made an excuse in many quarters for seriously interfering with the labour of aged miners ... because it is said or surmised that old men would be less able to avoid danger, and thus become more liable to accidents in mines'.[119] Although the 1880 Employers' Liability Act was limited in its scope and had little immediate effect (because mining employers frequently reached agreements with their workers to contract out of the legislation – especially in areas like south Wales where accident rates were high) it encouraged unions to push for further reform.[120] In 1897 the Workmen's Compensation Act introduced no-fault compensation that applied to any accident in the workplace irrespective of who was to blame.

The 1897 Act opened up the prospect of compensation for more disabled workers than ever before. Fighting compensation claims became an increasingly important part of trade union business in the early twentieth century. But the expansion of compensation would affect disabled miners' employment prospects in ways that seemed to confirm Rymer's fears about exclusion. Insurance schemes set up after 1897 to cover employers for their potential losses in compensation claims, sometimes refused to cover disabled, older or generally weaker workers, which made employers less likely to take the 'risk' of employing such personnel.[121] Ben Curtis and Steven Thompson have shown that very soon after the Workmen's Compensation Act came into force in July 1898, coal owners in south Wales dismissed as many as 1,000 older mineworkers on the pretext that they were especially prone to accidents. While the scale of such dismissals was exceptional, and probably reflective of the sour industrial relations in the south Wales coalfield during the prolonged strike of that year, similar occurrences happened elsewhere.[122] Witnesses to a Home Office Departmental Committee on the operation of Workmen's Compensation in 1903 reported that 'weak', sick, old and maimed workers were widely regarded by both employers and workmen's representatives as being more prone to accidents and slower to recover from them, making employers reluctant to hire them. The committee's report, published in 1904, concluded that older men and partially disabled workers were finding it harder to find and retain employment as a result of the Workmen's Compensation Act.[123] In the United States, where workmen's compensation laws had been established in an overwhelming majority of states by the early 1920s, the expense of insuring aged and partially incapacitated workers led to significant discrimination against such people in employers' hiring strategies, leading Rose to conclude that workmen's compensation represented a formative moment in the exclusion of disabled people from the workforce. While policy changes were by no means the only exclusionary force acting against the employment prospects of impaired people, she argues, they played a major role in forcing their marginalisation in the labour market.[124]

In early twentieth-century Britain, the question of whether compensation law, and other measures intended to benefit workers, changed the employment prospects of partially disabled workers, was more contested. For all those who argued that the laws worked against the interests of older or disabled employees, there were others in the labour movement, such as MP Keir Hardie, who maintained that compensation had little impact and it was in fact changes in technology or productivity norms, including a powerful rhetoric of worker efficiency, that served to exclude the less 'able'. Lack of consensus on the issue made the TUC unwilling to push for greater protection for these workers against discrimination.[125] Indeed, in the wake of workers' compensation legislation, some mines continued to provide light work for injured miners, possibly in an effort to keep down compensation costs under the new laws. During the economic troubles of the inter-war period, however, such practices became harder to maintain.[126] Ultimately, if the Employers' Liability Act and subsequent compensation legislation did not result in a wholesale displacement of disabled people from the world of work, these laws created a new rationale for disability discrimination that some employers found sufficiently persuasive to act upon. In an era of compensation, disabled workers faced renewed negative reflection on their capabilities, reliability and worth that saw them as posing a greater risk to their employers, their fellow workers and themselves.

Notes

1 'Capital and Labour', *Punch*, v (1843), 49. For a discussion of this image and its broader context, see Celina Fox, 'The Development of Social Reportage in English Periodical Illustration During the 1840s and early 1850s', *Past and Present*, 74:1 (1977), 90–111, especially 97. Matthias Reiss, 'The Image of the Poor and the Unemployed: the Example of *Punch*, 1841–1939' in Andreas Gestrich, Steven King and Lutz Raphael (eds), *Being Poor in Modern Europe: Historical Perspectives 1800–1940* (Oxford and Bern: Peter Lang, 2006), 389–415.
2 *The Times*, 29 April 1842, cited in Fox, 'Development of Social Reportage', 95.
3 *Hansard*, HC Deb 7 June 1842, vol. 63, cols 1320–53.
4 5 & 6 Victoria Cap. XCIX *An Act to Prohibit the Employment of Women and Girls in Mines and Collieries, to Regulate the Employment of Boys, and to Make Other Provisions Relating to Persons Working Therein*, 10 August 1842; Angela V. John, *By the Sweat of Their Brow: Women Workers at Victorian Coal Mines* (London: Routledge and Kegan Paul, 1984); cf. Catherine Mills, *Regulating Health and Safety in the British Mining Industries 1800–1914* (Farnham: Ashgate, 2010), 60.
5 5 & 6 Victoria Cap. XCIX.
6 D. G. Paz, 'Tremenheere, Hugh Seymour (1804–1893)', *Oxford Dictionary of National Biography*, Oxford University Press, 2004; online ed. January 2008, http://www.oxforddnb.com/view/article/27695, accessed 16 April 2015.
7 13 & 14 Victoria Cap. C., *An Act for Inspection of Coal Mines in Great Britain*, 14 August 1850; Mills, *Regulating Health and Safety*, chs 2 and 3. For subsequent renewals of the 1850 Act, see 18 & 19 Victoria Cap. CVIII, *An Act to Amend the Law for the Inspection of Coal Mines in Great Britain*, 14 August 1855; 23 & 24 Victoria Cap. CLI, *An Act for the Regulation and Inspection of Mines*, 28 August 1860; 25 & 25 Victoria Cap. LXXIX, *An Act to Amend the Law Relating to Coal Mines*, 7 August 1862.
8 Mills, *Regulating Health and Safety*; Raymond Challinor and Brian Ripley, *The Miners' Association: A Trade Union in the Age of the Chartists* (London: Lawrence and Wishart, 1968), 209–29.
9 Steve Sturdy, 'The Industrial Body' in Roger Cooter and John Pickstone (eds), *Companion to Medicine in the Twentieth Century* (London and New York: Routledge, 2003), 218; John Williams-Searle, 'Courting Risk: Disability, Masculinity, and Liability on Iowa's Railroads, 1868–1910', *The Annals of Iowa*, 58 (1999), 37; Jamie L. Bronstein, *Caught in the Machinery: Workplace Accidents and Injured Workers in Nineteenth-Century Britain* (Stanford, CA: Stanford University Press, 2008), 29.
10 Sturdy, 'The Industrial Body', 220–1.
11 35 & 36 Victoria Cap. LXXVI, *An Act to Consolidate and Amend the Acts Relating to the Regulation for Coal Mines and Certain Other Mines*, 10 August 1872.
12 Sarah F. Rose, '"Crippled" Hands: Disability in Labor and Working Class History', *Labor*, 2:1 (2005), 30, 41.

13 John Benson, *British Coalminers in the Nineteenth Century: A Social History* (Dublin: Gill and Macmillan, 1980), 197; Roy Church, *The History of the British Coal Industry Vol. 3: 1830–1913: Victorian Pre-Eminence* (Oxford: Clarendon Press, 1986), 674.
14 Alan B. Campbell, *The Lanarkshire Miners: A Social History of Their Unions, 1775–1874* (Edinburgh: John Donald, 1979), 62.
15 John Benson, 'English Coal-Miners' Trade-Union Accident Funds, 1850–1900', *Economic History Review* 28:3 (1975), 401–12; James A. Jaffe, 'The State, Capital, and Workers' Control during the Industrial Revolution: The Rise and Fall of the North-East Pitmen's Union, 1831–2', *Journal of Social History*, 21 (1988), 728.
16 Benson, 'English Coal-Miners' Trade-Union Accident Funds', 405–7.
17 TNA, FS 28/8 Durham Miners' Association, 1872–1914, Rule book.
18 National Archives of Scotland, FS7/1, *Rules of the Larkhall Miners' Mutual Protection, Accident and Funeral Association*, 8 December 1874, 2, 4.
19 Bronstein, *Caught in the Machinery*, 120.
20 John Benson, 'Coalminers, Accidents and Insurance in Late Nineteenth-Century England', in Bernard Harris (ed.), *Welfare and Old Age in Europe and North America: The Development of Social Insurance* (London: Pickering & Chatto, 2012), 22.
21 Northumberland Archives, 2/DE.4.6/39, Letters from John Crooks, 1788–96, Hartley Colliery Strike and Payment to Injured Miners, 1793.
22 Indeed, as Benson notes, smart money was unique to north-east England and 'did not exist elsewhere' in the country. Benson, 'Coalminers, Accidents and Insurance', 22.
23 United Association of Colliers, *A Voice from the Coal Mines, Or, a Plain Statement of the Various Grievances of the Pitmen of the Tyne and Wear: Addressed to the Coal Owners, Their Head Agents, and a Sympathizing Public* (South Shields: J. Clark, 1825), 28.
24 *Lloyd's Weekly Newspaper*, 23 June 1844, 5.
25 Ibid; *The Times*, 22 May 1844; 'The Pitmen's Strike', *The Newcastle Courant*, 10 May 1844.
26 United Association of Colliers, *Voice from the Coal Mines*, 9, 16, 21.
27 Ibid., 9.
28 PP 1849 [1109], *Report of the Commissioner Appointed, Under the Provisions of the Act 5 & 6 Vict., c. 99, to Inquire into the Operation of that Act, and into the State of the Population in the Mining Districts, 1849*, 16.
29 Bronstein, *Caught in the Machinery*, 147.
30 Marjorie Levine-Clark, '"The Entombment of Thomas Shaw:" Mining Accidents and the Politics of Workers' Bodies', *Victorian Review*, 40:2 (2014), 23.
31 Challinor and Ripley, *The Miners' Association*, 209–29; Mills, *Regulating Health and Safety*, 103; A. J. Taylor, 'The Miners' Association of Great Britain and Ireland, 1842–48: A Study in the Problem of Integration', *Economica*, 22:85 (1955), 59.

32 Jamie L. Bronstein, 'The Hartley Colliery Disaster', *Victorian Review*, 40:2 (2014), 11. The petition, combined with sympathetic public opinion, helped pressure MPs to introduce a change to the law in June 1862 to require collieries to have two shafts.
33 Challinor and Ripley, *The Miners' Association*, 128.
34 Frederick [sic] Engels, *The Condition of the Working Class in England* (London: Panther, 1969), 271, 274, 278.
35 *The British Miner and General Newsman*, n.s. no. 1, 7 March 1863. Bronstein, *Caught in the Machinery*, 79–80.
36 Ibid., no 5, 11 October 1862, 4.
37 Ibid., n.s., no. 13, 30 May 1863, 3.
38 Rose, '"Crippled" Hands', 47; David T. Mitchell and Sharon Snyder, *Narrative Prosthesis: Disability and the Dependencies of Discourse* (Ann Arbor: University of Michigan Press, 2000), 47–64.
39 Church, *British Coal Industry*, 675.
40 Benson, *British Coalminers*, 189–90; J. H. Morris and L. J. Williams, *The South Wales Coal Industry, 1841–1875* (Cardiff: University of Wales Press, 1958), 270; Mills, *Regulating Health and Safety*, 114–15.
41 Chris Wrigley, 'Unions', in Joel Mokyr (ed.), *The Oxford Encyclopaedia of Economic History*, Vol. 5 (Oxford & New York: Oxford University Press, 2003), 154.
42 See also Dick Geary, 'The Myth of the Radical Miner' in Stefan Berger, Andy Croll and Norman Laporte (eds), *Towards a Comparative History of Coalfield Societies* (Aldershot: Ashgate, 2005), 43–64.
43 *Colliery Guardian*, 5 October 1867, quoted in Huw Beynon and Trevor Austrin, *Masters and Servants: Class and Patronage in the Making of a Labour Organisation. The Durham Miners and the English Political Tradition* (London: Rivers Oram, 1984), 31.
44 PP 1877 (285), *Report from the Select Committee on Employers Liability for Injuries to their Servants; Together with the Proceedings of the Committee, Minutes of Evidence, and Appendix*, 15.
45 John Williams-Searle, 'Cold Charity: Manhood, Brotherhood and the Transformation of Disability, 1870–1900', in Paul K. Longmore and Lauri Umansky (eds), *The New Disability History: American Perspectives* (New York: New York University Press, 2001), 157–86 (quote from 160).
46 John Rule, *The Labouring Classes in Early Industrial England, 1750–1850* (London: Longman, 1986), 313–15, 331–40; Robert Duncan, *The Mineworkers* (Edinburgh: Birlinn, 2005), 133; Morris and Williams, *South Wales Coal Industry*, 274–84.
47 Rose, '"Crippled" Hands', 44.
48 Alan Campbell, for example, notes the frequent violence in Scottish mining strikes: *Lanarkshire Miners*, 77.
49 TNA, PC 1/18/19, Affidavit of Thomas Barnes, 20 March 1789.
50 *The Times*, 8 November 1842.

51 For other examples of people injured in violence connected to industrial conflict in the coalfields, see: *The Times*, 9 May 1832; ibid., 2 August 1832.
52 TNA, HO 52/19, 216–217; David J. V. Jones, *Before Rebecca: Popular Protests in Wales, 1793–1835* (London: Allen Lane, 1973), 86–113.
53 Edward Rymer, *The Martyrdom of the Mine, Or, A 60 Years Struggle for Life* (Middlesbrough, 1898), 8.
54 PP 1846 [737], *Report of the Commissioner Appointed Under the Provisions of the Act 5 & 6 Vict., c. 99, to Inquire into the Operation of that Act, and into the State of the Population in the Mining Districts, 1846*, 15, 9.
55 Alexander Dalziel, *The Colliers' Strike in South Wales* (Cardiff: The Western Mail Offices, 1872), 171.
56 Andy Croll, 'Starving Strikers and the Limits of the "Humanitarian Discovery of Hunger" in Late Victorian Britain', *International Review of Social History*, 56 (2011), 104.
57 *The Times*, 20 April 1827.
58 Ibid., 3 October 1844.
59 Robert Colls, *Pitmen of the Northern Coalfield: Work, Culture, and Protest, 1790–1850* (Manchester: Manchester University Press, 1987), 15.
60 For example, Campbell, *Lanarkshire Miners*; Rule, *The Labouring Classes*, 338–40; Morris and Williams, *South Wales Coal Industry*, 262–4.
61 TNA, HO 52/19, 216–17.
62 Richard Fynes, *The Miners of Northumberland and Durham: A History of Their Social and Political Progress* (Blyth: John Robinson, 1873), 111.
63 PP 1846 [737], 16
64 Campbell, *Lanarkshire Miners*, 51, 108–9, 264–6; Colls, *Pitmen*, 30.
65 PP 1844 [592], *Report of the Commissioner Appointed Under the Provisions of the Act 5 & 6 Vict. c. 99, to Inquire into the Operation of that Act, and into the State of the Population in the Mining Districts*, 32.
66 PP 1845 [670], *Report of the Commissioner Appointed Under the Provisions of the Act 5 & 6 Vict. c. 99, to Inquire into the Operation of that Act, and into the State of the Population in the Mining Districts, 1845*, 11; PP 1856 [2125], *Report of the Commissioner Appointed Under the Provisions of the Act 5 & 6 Vict. c. 99, to Inquire into the Operation of that Act, and into the State of the Population in the Mining Districts, 1856*, 42.
67 Colls, *Pitmen*, 32.
68 John Belchem, *Industrialization and the Working Class: The English Experience, 1750–1900* (Aldershot: Scolar Press, 1990), 16–17.
69 Duncan, *The Mineworkers*, 134; Challinor and Ripley, *The Miners' Association*, 137.
70 Glamorgan Archives UM/1/18, Merthyr Tydfil Board of Guardians Minutes, 16 March 1875. The legal right to relief for strikers is examined in Andy Croll, 'Strikers and the Right to Poor Relief in Victorian Britain: The Making of the Merthyr Tydfil Judgment of 1900', *Journal of British Studies*, 52 (2013), 128–52.

71 Bronstein, *Caught in the Machinery*, 35, 46.
72 Benson, *British Coalminers*, 179.
73 Fynes, *Miners of Northumberland and Durham*, 139–40.
74 For example, Durham Record Office, D/DCOA 72, Durham Coal Owners' Association, Proceedings of Joint Committee, 30 September 1880.
75 Benson, 'English Coal-Miners' Trade-Union Accident Funds', 409–11.
76 Scrutator, *An Impartial Enquiry into the Existing Causes of Dispute between the Coal Owners of the Wear and Tyne and Their Late Pitmen* (Houghton-Le-Spring: Printed for the author by J. Beckwith, 1832), 6. For similar allegations about the strike made by owners themselves: Coalowners of the Rivers Tyne and Wear, *Report by the Committee of the Coalowners Respecting the Present Situation of the Trade* (Newcastle: Printed by W., E., and H. Mitchell, Newcastle, 1832), 6–7.
77 PP 1846 [737], 8; Jaffe, 'The State, Capital, and Workers' Control', 721.
78 'Articles of the Govan Colliery Friendly and Free Labour Society', PP 1851 [1422], *Coal Mines. Reports of Messrs. Dunn, Dickinson, and Morton, Inspectors of Coal Mines, to Her Majesty's Secretary of State*, 18–22. PP 1856 [2132], *Report of the Commissioner Appointed under the Provisions of the Act 5 & 6 Vict. c. 99, to Inquire into the Operation of that Act, and into the State of the Population in the Mining Districts, 1856*, 37–41, 53–58; PP 1842 (381), *Appendix to the First Report of the Commissioners. Mines. Part 1. Reports and Evidence from Sub-Commissioners*, 356–9; Campbell, *Lanarkshire Miners*, 71; TNA, FS 1/120, 'Rules of the Hetton Colliery Agents and Workmen's Friendly Society', [Rule 3].
79 Campbell, *Lanarkshire Miners*, 71.
80 Benson, *British Coalminers*, 104–5.
81 Fynes, *Miners of Northumberland and Durham*, 74–5.
82 Challinor and Ripley, *The Miners' Association*, 135.
83 Ibid., 80.
84 TNA, HO 45/7692 Home Office: Registered Papers. Disturbances: Evictions at Cramlington Colliery, Northumberland (1865).
85 Fynes, *Miners of Northumberland and Durham*, 74.
86 Cf. Croll, 'Starving Strikers'.
87 PP 1843 [508], *Midland Mining Commission. First Report. South Staffordshire*, xxv. We are grateful to Jamie Bronstein for encouraging us to think about this possibility.
88 13 & 14 Victoria Cap. C, Section VIII.
89 Bartrip and Burman, *Wounded Soldiers of Industry*, ch. 4.
90 P. W. J. Bartrip, *Workmen's Compensation in Twentieth-Century Britain: Law, History and Social Policy* (Aldershot: Gower, 1987), 5.
91 TNA, HO 45/5547, Memorial of Northumberland and Durham Miners to the Home Secretary, 16 December 1854. Our emphasis.
92 Bartrip and Burman, *Wounded Soldiers of Industry*, 110.
93 Bronstein, *Caught in the Machinery*, 146.
94 Ibid., 103.

95 *Paisley Herald and Renfrewshire Advertiser*, 4 August 1855.
96 'Damages from a Colliery Owner for Personal Injury', *Colliery Guardian and Journal of the Coal and Iron Trades*, 3 April 1858; Bronstein, *Caught in the Machinery*, ch. 5. On compensation and the railways see Audrey C. Giles, 'Railway Accidents and Nineteenth-Century Legislation: "Misconduct, Want of Caution or Causes beyond their Control"', *Labour History Review*, 76:2 (2011), 121–42; Michael Quick, 'Mid-Victorian Compensation Culture', *Journal of the Railway and Canal Historical Society*, 192 (2005), 110–17.
97 *Griffiths v. Gidlow* in E. T. Hurlstone and J. P. Norman, *The Exchequer Reports: Reports of Cases Argued and Determined in the Courts of Exchequer & Exchequer Chamber*, vol. III (London: H. Sweet, W. Maxwell ,V. & R. Stevens and G. S. Norton, 1859), 648–56.
98 Bartrip and Burman, *Wounded Soldiers of Industry*, 96.
99 35 & 36 Victoria Cap. LXXVI, Section 68.
100 Bronstein, *Caught in the Machinery*, 152–4.
101 Benson, *British Coalminers*, 200.
102 Ibid., 155.
103 The fullest accounts can be found in Bartrip and Burman, *Wounded Soldiers of Industry*, ch. 5; Bronstein, *Caught in the Machinery*, ch. 5.
104 *Northern Echo*, 22 May 1876.
105 'Compensation for Injuries Bill', *Colliery Guardian*, 27 December 1878.
106 *Hansard*, HC Deb 21 June 1878, vol. 241, cols 67–98. For more on the Keelmen's scheme, see P. H. J. H. Gosden, *The Friendly Societies in England 1815–1875* (Manchester: Manchester University Press, 1961), 6.
107 Letter from Joseph S. Pease, *The Times*, 22 July 1880.
108 *Northern Echo*, 19 July 1880.
109 Ibid.
110 'The Durham Miners' Demonstration', *Northern Echo*, 2 August 1880.
111 43 & 44 Victoria Cap. 42. *An Act to Extend and Regulate the Liability of Employers to Make Compensation for Personal Injuries Suffered by Workmen in Their Service*, 7 September 1880; Bronstein, *Caught in the Machinery*, 164.
112 Bartrip and Burman, *Wounded Soldiers of Industry*, 161–4.
113 *Glasgow Herald*, 30 December 1880; Bartrip and Burman, *Wounded Soldiers of Industry*, 161.
114 Ibid., 164.
115 Durham Record Office, D/DCOA 72 Durham Coal Owners' Association, Proceedings of the Joint Committee, 1872–81.
116 Church, *British Coal Industry*, 584; cf. Benson, *British Coalminers*, 43.
117 35 & 36 Victoria Cap. LXXVI, Section 52.
118 Campbell, *Lanarkshire Miners*, 107; Colls, *Pitmen*, 37–8.
119 Rymer, *Martyrdom of the Mine*, 27.
120 Ben Curtis and Steven Thompson, '"This is the Country of Premature Old Men:" Ageing and Aged Miners in the South Wales Coalfield, c. 1880–1947', *Cultural*

and *Social History*, 12:4 (2015), 598; Bartrip and Burman, *Wounded Soldiers of Industry*, 173.
121 Bartrip and Burman, *Wounded Soldiers of Industry*, 212.
122 Curtis and Thompson, 'This is the Country of Premature Old Men', 597; Durham Record Office, D/X 1005/21, correspondence to let aged or infirm men go at Chopwell Colliery, 10 September 1912.
123 Bartrip, *Workmen's Compensation*, 43.
124 Nate Holdren, 'Incentivizing Safety and Discrimination: Employment Risks under Workmen's Compensation in the Early Twentieth Century United States', *Enterprise and Society*, 15 (2014), 31–67; Sarah F. Rose, *No Right To Be Idle: The Invention Of Disability, 1840s–1930s* (Chapel Hill: University of North Carolina Press, 2017), 161–71, 224.
125 Bartrip, *Workmen's Compensation*, 52, 72; Sturdy, 'The Industrial Body'.
126 Mike Mantin, 'Coalmining and the National Scheme for Disabled Ex-Servicemen after the First World War', *Social History*, 41:2 (2016), 161.

CONCLUSION

The Industrial Revolution produced injury, illness and disablement on a large scale and nowhere was this more visible than in coalmining. While the loss of lives in large-scale mining disasters is still commemorated today, and forms part of the cultural memory of coalmining in areas where pits have long since closed down, there are no memorials to the many thousands who were disabled in the industry.[1] Yet the experiences of those whose bones were broken, whose bodies were crushed, 'lamed' or maimed, or who entered old age prematurely as a result of being 'worn out' by their labours or by the shortness of breath brought on by lung disease, matter just as much to mining's history as those who lost their lives. Some were rendered incapable of work, either permanently or temporarily. Others 'worked through' chronic illness or impairment, or took up other roles within their communities. The incidence of injury and impairment in coalmining and other dangerous trades led to a range of medical, welfare and political responses, some of which have left a lasting legacy. These include the statutory regulation of workplace health and safety, the principle, albeit contested, that employers should bear some responsibility for accidents at work and the belief that the welfare needs of disabled people differ from those of the general poor and unemployed.

Disability was essential to the Industrial Revolution, but historical experiences of disability are far more complex than previously argued. Historical materialist accounts have emphasised that the advent of industrial capitalism led to the marginalisation of disabled people as economically unproductive 'burdens', whose inability to conform to more stringent productivity demands, work or time discipline meant that they could no longer compete in the workplace. Yet the coal industry during its period of rapid expansion between 1780 and 1880 presents a more complicated picture. On the one hand, the idea that coalminers were a 'picked' body of workers probably meant that people with certain congenital impairments or 'weak' constitutions had long been excluded from mine work, although such exclusion was never universal. On the other hand, if British coalminers were admired for their physical prowess, the acquired diseases and injuries associated with their toils meant that many experienced some degree of impairment. Our evidence shows that rather than leaving the world of work, these 'disabled' miners were expected to return to productive employment if capable of doing so. Such workers were valued for their skills and experience, even more so when labour was scarce, such as during strikes. For much of our period, elements of the 'somatic flexibility' believed to have enabled disabled people to remain economically productive in the 'pre-industrial' era remained intact. The practice of working in family groups relatively free from supervision, for instance, continued at many mines throughout the nineteenth century. Combined with customs such as piecework and worker-controlled restriction of output, such practices enabled some 'disabled' miners to remain active in the workforce.

Despite labour historians' acknowledgement that impairment has been a common consequence of work in the past, disabled people rarely appear as *workers* in the histories they have written.[2] This absence 'naturalises' the idea that disabled bodies are, to borrow Ava Baron's phrase about the gender bias of labour history, 'out of place' in the world of work.[3] If, on the one hand, the risks to health from accident or disease increased as a result of coalmining's expansion, potentially leading to greater disablement, on the other hand workers with impairments also contributed to the expanding coal industry through their own labour. In this respect, disabled mineworkers helped make the Industrial Revolution. Some of this work was relatively low status and poorly remunerated. Putting adult men made 'cripples' in mine accidents to 'boys' work' could be demeaning in an industry where the status of work related to hierarchies of age. However, we should not assume that the work done by 'cripples' was automatically devalued. Such work could still invoke feelings of pride or represent an important contribution to an individual or family's efforts to 'make ends meet'. As feminist historians have taught us, just as we should not dismiss or devalue women's work in the past simply because it was considered 'lowly', neither should we do the same with the labour disabled people have performed.[4] Rather than judging disabled people's relationship with work in the past simply in terms of their 'inclusion' or 'exclusion', we need to pay attention to the meanings and value of their work within particular occupational or familial contexts.

Accounts of seriously injured miners returning to work are appealing because they show the economic productivity of disabled people in the past. Yet this does not mean collieries, or pit villages, were free of prejudice, or that it was always easy for disabled mineworkers to return to work or make a living. A person's experiences depended on many factors – from geological conditions and their own physical capacities and skills to the attitudes of employers and neighbours alike. In a society where moral assessments of causes of impairment were as important as its consequences in determining sympathy and support, responses to disablement could vary widely. Mining communities might come together to help disabled men find new employment, but could be hostile to strangers with impairments, particularly beggars. Similarly, unions might support disabled miners, but drew a line at helping those whose behaviour had endangered themselves or their workmates.[5]

Undoubtedly those who were reliant on poor relief, or payments from friendly societies, were at risk of impoverishment as these benefits were often far less than what they could earn, especially if they had previously been employed in lucrative jobs such as coal-cutting. In early nineteenth-century Scotland in particular, women's employment as coal bearers had made an important contribution to the income of households where men were incapacitated, and the banning of females from working underground in 1842 robbed these families of this source of support. Women's unpaid work in supporting their husbands and sons by washing clothes, carrying water for baths and keeping house made an important – if seldom acknowledged – contribution to the domestic economy, and the loss of miners' wives' labour through incapacity could also place pressure on households. In communities where the productivity of the adult male breadwinner was considered particularly important – symbolically as well as economically – some men undoubtedly viewed incapability for paid work with despondency.

While disability might modify family dynamics in mining communities and challenge men's status, emasculation was never inevitable – not even for miners unable to maintain main breadwinner status. There were many options available to disabled men to prove their manliness beyond the world of work. Some injured mineworkers appropriated the old image of miners as tough, hard drinking and prone to violence to assert their strength, while others embraced new opportunities for spiritual leadership through evangelical nonconformity. Still more insulated themselves, to a degree, from the emasculating potential of disability through marriage and fatherhood. Although women might, on occasion, view suitors with impairments in a negative light, there is little evidence that injury or chronic illness in the coalfields was ever an absolute barrier to the formation or maintenance of enduring and meaningful relationships. Neither did it deprive miners of the ability to express themselves as men sexually – a fact amply demonstrated by the many children fathered by disabled colliers long after the onset of impairment, but often forgotten in history books. As husbands, fathers and lovers, then, it was hard for disabled miners to be *completely* unmanned by work-related incapacity, either in their own eyes or in others.

Disablement may also have provided an opportunity for some to escape a dangerous and physically demanding workplace. Two fifteen-year-old boys, Morgan Thomas and Giles Giles, who lost limbs at Hirwaun Colliery near Merthyr Tydfil, told the 1842 Children's Employment Commission that during their lay off from work they had been able to go to school and learn to write. At a time when educational opportunities were scarce for youngsters who were under pressure to go to work at an early age, such experiences were valuable despite their circumstances. Though both lads saw their future as uncertain, the acquisition of literacy was intended to provide opportunities to earn a living in clerical work.[6]

Over time, some of the 'somatic flexibility' that had enabled disabled mineworkers to work in collieries was eroded. The spread of longwall mining and the increasing reliance on winding machinery as the demand for coal grew and deep mining expanded, for instance, reduced the autonomy of mineworkers to determine their own work routines and rhythms. Legal changes were also significant. 'Special rules' and increased supervision of workers intended to improve safety at mines similarly affected miners' ability to choose how they worked, while employers' liability and compensation laws designed to protect injured employees laid the basis for greater discrimination against 'at risk' workers in the future. Yet we should not exaggerate the impact of these changes. Coalmining remained an industry receptive to the re-employment of men after serious injury well into the twentieth century, despite the downturn in the sector's economic fortunes between the First and Second World Wars.[7]

The development of industrial society is also important to the history of medical and welfare responses to disablement. The dangers of mine work meant that British coalminers were among the first occupational groups to receive dedicated medical care, first via surgeons funded through colliery 'sick clubs' and later on through specialist institutions aimed at supporting recuperation. During this period, mining communities were increasingly viewed as profitable medical markets, for both orthodox and irregular practitioners. Although access to medical services was uneven, providing medical care was a means for employers to demonstrate both their paternalistic concern for their employees and to discipline them. Medical experts played an increasingly important role as the gatekeepers to welfare services, but their authority was sometimes challenged. Many diseased and disabled mineworkers sought independence and negotiation in their dealings with doctors, and some decisions, whether to dissect the bodies of those who had succumbed to lung diseases or to amputate damaged limbs, were opposed outright by miners and their families. Therefore, although the injuries and diseases of coalminers were subject to much medical theorising in this period, the process of 'medicalisation' was far from smooth. Right up to the end of our period, the authority of medical men in diagnosing conditions or prescribing a course of treatment was never absolute.[8]

As well as increasing the demand for medical services within the coalfields, the large number of mining accidents also put pressure on existing welfare resources such as the Poor Law, stimulating new responses. While friendly societies flourished in coalmining areas, some were reluctant to admit those working in such a dangerous occupation lest they bankrupt schemes based on shaky actuarial foundations. This problem was addressed by the amalgamation of societies into affiliated orders, but mineworkers were sometimes asked to pay higher subscriptions. Friendly societies and trade union accident funds provided important assistance to injured mineworkers, but only on a relatively short-term basis and with payments often decreasing over time. From 1862 the Northumberland and Durham Permanent Relief Fund established a new model of relief, which recognised that disabled members needed lasting support rather than the diminishing returns of friendly society payments. This model soon spread to other mining districts and was well entrenched in most English coalfields by the end of our period.[9] While miners continued to receive support from a variety of sources, sometimes simultaneously, the permanent relief fund movement symbolically represented a significant staging post in the development of the modern idea of disability as a long-term, permanent state of incapacity that required dedicated welfare responses.[10]

The physical toll of coalmining affected how miners were viewed by others and how they saw themselves. It was common for middle-class commentators to describe coalminers as a 'distinct race of men', whose working conditions produced certain physical and intellectual characteristics and whose exposure to accident affected their social and moral 'habits'. Occupationally specific deformities were viewed as 'trade marks' that delineated coalminers and other groups of workers in industrialising Britain, demonstrating the importance of the body in social classification. Within mining communities, the shared risk of death or disablement fostered a strong sense of mutualism and helped to shape working-class identity.[11] The figure of the disabled miner was a powerful rhetorical tool for trade unionists, used to support their campaigns for better safety and provision for the injured, as well as to stand for the sufferings of all coalminers in the face of what mining activist Edward Rymer (himself 'half blind' and a 'cripple') termed the 'mighty Juggernaut' of industrial capitalism.[12] The perceived 'victimhood' of disabled people could be a powerful resource in highlighting the cruel practices of employers during disputes, as the stories of elderly and impaired persons being evicted from company housing during the 1844 strike in north-east England illustrated. Yet disabled miners were involved more proactively in industrial politics, sometimes independently of trade unions, which might not always sympathise with their cause. Breaking terms of employment to escape unhealthy or dangerous collieries, seeking to supply shortages of labour during strikes, challenging colliery doctors, or fighting for compensation in the courtroom were all means by which disabled miners asserted themselves politically during this period.

While this book has opened up new perspectives on disability in Britain's coalfields, it points to the need for further research. First, more work is needed to compare the experiences of coalminers with those disabled in other important economic sectors such as textiles, metallurgy, transport and agriculture. If the medical and welfare needs of other British disabled workers were similar to those of miners, how did their different occupational backgrounds and cultures affect their experiences of care, of work, of family and how others treated and perceived them during the Industrial Revolution? By exploring other sectors of the industrialising economy, and by comparing further the experiences of workers in the industrial areas of Europe, North America, and the Majority World, a more nuanced picture still of disability and industrialisation will emerge.[13] Second, more research is needed on gendered understandings and experiences of disablement in relation to work. Even in areas where women were employed underground before 1842, mining was considered a 'masculine' occupation, and it is men's experiences of disability that were foregrounded in our sources. However, when female Scottish coal bearers complained to the 1842 Children's Employment Commission of their ability to bear children being affected by their bodily toil, they presented different fears about disablement to those expressed by their husbands and sons. When disability is seen as inability to work, it is often the ability to do *paid* work that is had in mind. Opening up definitions of work to include the numerous unpaid tasks traditionally performed by women, from the emotional labour of caring to domestic chores, will allow new perspectives on the relationship between disability and work to emerge.

Finally, it is likely that the experiences of those disabled in coalmining, an industry valued by contemporaries for its vital role in fuelling industrial expansion, differed from those disabled people whose economic activities took place within asylums or sheltered workshops.[14] Trade unionists and others, while often emphasising the suffering of disabled miners also claimed status for them, seeing them as deserving the same level of recognition as military veterans who sacrificed their limbs in battle.[15] Such honorific – and deeply gendered – claims were rarely made for disabled people working in institutional settings.[16] Previously able-bodied workers who experienced impairment through bodily 'sacrifice' might be able to claim a sense of heroism or worth that was not as easily accessible for those with congenital impairments or who had become disabled before starting work. Future research needs to explore more closely the varying experience of work between different disabled populations rather than focusing simply on the 'disabled'/'able-bodied' dichotomy.

Analysing historical experiences of disability and work reminds us that disabled people have always worked in the past whenever they had the opportunity or ability to do so.[17] We need to recognise and value the economic contributions of disabled people, but not use the past to make unrealistic policies or demands in the present. Finding people with crutches and wooden legs in Britain's nineteenth-century coal mines shatters preconceptions that such work was the sole preserve of the 'able-bodied', but their history is not intended to be 'inspirational'. Their presence reflects more the struggle for survival and the inadequacies of other sources of support than it does economic empowerment. And while aspects of Victorian paternalism may look attractive in the context of the increasing casualisation of labour in twenty-first-century Britain that has left many without access to sick pay, it would be anachronistic to use it as evidence of a more 'positive' attitude towards disabled people's employment in the past.[18] Those who made the journey from pithead to sickbed and beyond faced significant challenges but their story is crucial to understanding our industrial past.

Notes

1 'Felling Pit Disaster Remembered on 200th Anniversary', http://www.bbc.co.uk/news/uk-england-tyne-18194972, accessed 26 March 2017.
2 Sarah F. Rose, '"Crippled Hands: Disability in Labor and Working Class History', *Labor*, 2:1 (2005), 45–7.
3 Ava Baron, 'Masculinity, the Embodied Male Worker, and the Historians' Gaze', *International Labor and Working-Class History*, 69 (2006), 154.
4 Joanna Bourke, 'Housewifery in Working-Class England 1860–1914', *Past and Present*, 143 (1994), 163–97.
5 For a comparative case study, see John Williams-Searle, 'Cold Charity: Manhood, Brotherhood and the Transformation of Disability, 1870–1900', in Paul K. Longmore and Lauri Umansky (eds), *The New Disability History: American Perspectives* (New York: New York University Press, 2001), 157–86.
6 PP 1842 (382), *Children's Employment Commission. Appendix to the First Report of Commissioners. Mines. Part II*, 552.
7 Kirsti Bohata, Alexandra Jones, Mike Mantin and Steven Thompson, *Disability in*

Industrial Britain: A Cultural History of Illness, Injury and Impairment in the Coal Industry, 1880–1948 (Manchester: Manchester University Press, forthcoming).

8 See also James C. Riley, *Sick Not Dead: the Health of British Workingmen during the Mortality Decline* (Baltimore, MD and London: Johns Hopkins University Press, 1997), ch. 3.

9 George L. Campbell, *Miners' Insurance Funds: their Origins and Extent* (London: Waterlow and Sons, 1880).

10 Roger Cooter, 'The Disabled Body' in Roger Cooter and John Pickstone (eds), *Companion to Medicine in the Twentieth Century* (London and New York: Routledge, 2003), 367–84.

11 See also Rose, '"Crippled" Hands', 41–2.

12 Edward Rymer, 'The Poor Miner', *Miner and Workmen's Advocate*, no. 112, 22 April 1865.

13 Daniel Blackie, 'Disability and Work during the Industrial Revolution in Britain', in Michael Rembis, Catherine Kudlick and Kim E. Nielsen (eds), *The Oxford Handbook of Disability History* (New York: Oxford University Press, forthcoming).

14 Sarah F. Rose, 'Work', in Rachel Adams, Benjamin Reiss and David Serlin (eds), *Keywords for Disability Studies* (New York and London: New York University Press, 2015), 188.

15 For example, 'Institution for Disabled Miners', *Cardiff and Merthyr Guardian*, 18 February 1837.

16 Cf. Gordon Phillips, *The Blind in British Society: Charity, State and Community, c. 1780–1930* (Aldershot: Ashgate, 2004), ch. 3.

17 Rose, 'Work', 187.

18 Emma Jacobs, 'The Gig Economy: Freedom From a Boss, or Just a Con?', *New Statesman*, 20 March 2017, http://www.newstatesman.com/politics/economy/2017/03/gig-economy-freedom-boss-or-just-con, accessed 28 March 2017.

Select bibliography

Primary sources

Manuscripts

Durham County Record Office
D/DCOA – Durham Coal Owners' Association Records, 1872–1947.

Glamorgan Archives
DODD – Oddfellows Society Records.
P62/4–7 – Llantrisant Civil Parish Records: Vestry and Parish Meeting Minute Books, 1771–1815 and 1861–97.
UM/1 – Merthyr Tydfil Poor Law Union, Minutes of Board of Guardians, 1836–1930.

Glasgow City Archives, Mitchell Library
CO1/22/44 – Blantyre Parochial Board, Register of Poor, 1845–64.
CO1/27 – Carluke Parochial Board/Parish Council Records, 1846–1935.
CO1/37 – Dalziel Parochial Board/Parish Council Records, 1858–1935

Glasgow University Archives
UGD1 – Govan Colliery Financial Records, 1849–1948.

Gwent Archives
CSWBGP/M1 – Pontypool Union Board of Guardians' Minutes, 1836–1930.
CSW/BGP/I/222–230 – Pontypool Union Workhouse Admissions and Discharges, 1858–1935.
D3293/A – Newport and Monmouthshire Hospital/Royal Gwent Hospital Annual Reports, 1854–1946.
D32.149 – Rules of a Friendly Society of Tradesmen and Others, Called The Star Friendly Society.

The National Archives, London
FS 1, FS 2 – Friendly Societies Rules and Amendments, Series I, c. 1784–75.
FS 5 – Friendly Societies, Branches, Rules and Amendments, etc., Series I, 1855–1912.
HO 17 – Home Office: Criminal Petitions, Series I, 1819–58.
HO 45 – Home Office: Registered Papers, 1839–1979.
HO 52 – Home Office: Counties Correspondence, 1820–50.
HO 73 – Home Office: Various Commissions: Records and Correspondence, 1786–1949.
HO 87 – Home Office: Factory and Mines Entry Books, 1836–1921.

HO 95 – Home Office: Mines Entry Books, 1855–71.
MH 2 – Poor Law Commission: Rough and Classified Minute Books, 1834–47.
MH 12 – Local Government Board and Predecessors: Correspondence with Poor Law Unions and Other Local Authorities, 1834–1900.
PC 1 – Privy Council and Privy Council Office: Miscellaneous Unbound Papers, 1481–1946.

National Archives of Scotland, Edinburgh.
FS 1 – Friendly Societies. 1st Series.
FS 4 – Friendly Societies. 3rd Series.

NHS Greater Glasgow and Clyde Archives, Glasgow University
HB/14 and HH/67 – Records of Glasgow Royal Infirmary, hospital, Glasgow, Scotland, 1787–1988.
HB/47 – Records of Glasgow Ophthalmic Institution, 1870–1965.

Northumberland Archives
2/DE.4.6/1–71 – Letters from John Crooks, 1788–96.

Richard Burton Archives, Swansea University
SWCC – South Wales Coalfield Collection.

Tyne and Wear Archives, Newcastle
CH.KH/1 – Keelmen's Hospital, Newcastle upon Tyne, Minutes, 7 January 1739/40 to 17 December 1842.
CH.MPR – Northumberland and Durham Miners Permanent Relief Fund Friendly Society Records.
PU.SS/1 – South Shields Poor Law Union: Board of Guardians, Minutes, 1836–1930.
S.PAM/1 – Society for the Prevention of Accidents in Coal Mines, Sunderland: Correspondence, Technical notes, Drawings and Pamphlets Accumulated by the Society's Secretary William Burn, 1802–38.

Newspapers and periodicals

Aberdare Times
Association Medical Journal
Birmingham Daily Post
The Bristol Mercury
British Medical Journal
The British Miner and General Newsman
The Cambrian
Cardiff and Merthyr Guardian
Cardiff Times

The Charter
Colliery Guardian and Journal of the Coal and Iron Trades
Daily Advertiser
Dean Forest Mercury
Edinburgh Medical Journal
The Examiner
The Foresters' Miscellany
The Gentleman's Magazine
Glasgow Herald
Household Words
Illustrated London News
Lloyd's Weekly Newspaper
London Chronicle
London Medical Gazette, or Journal of Practical Medicine
Medical Chirurgical Transactions
Merthyr Telegraph and General Advertiser for the Iron Districts of South Wales
Methodist Magazine
Miner and Workman's Advocate
Monmouthshire Merlin
Monthly Supplement of the Penny Magazine of the Society for the Diffusion of Useful Knowledge
The Morning Post
Newcastle Courant
Northern Echo
Paisley Herald and Renfrewshire Advertiser
Poor Law Unions' Gazette
Primitive Methodist Magazine
Provincial Medical Surgery Journal
Public Advertiser
Punch
St. James's Chronicle or the British Evening Post
Scottish Poor Law Magazine
South Wales Daily News
Sunday Chronicle
The Times
The Wesleyan-Methodist Magazine
Western Mail
Whitehall Evening Post

UK statutes

5 & 6 Victoria Cap. XCIX, *An Act to Prohibit the Employment of Women and Girls in Mines and Collieries, to Regulate the Employment of Boys, and to Make Other Provisions Relating to Persons Working Therein*, 10 August 1842.

13 & 14 Victoria Cap. C, *An Act for Inspection of Coal Mines in Great Britain*, 14 August 1850.
18 & 19 Victoria Cap. CVIII, *An Act to Amend the Law for the Inspection of Coal Mines in Great Britain*, 14 August 1855.
23 & 24 Victoria Cap. CLI, *An Act for the Regulation and Inspection of Mines*, 28 August 1860.
25 & 25 Victoria Cap. LXXIX, *An Act to Amend the Law Relating to Coal Mines*, 7 August 1862.
35 & 36 Victoria Cap. LXXVI, *An Act to Consolidate and Amend the Acts Relating to the Regulation of Coal Mines and Certain Other Mines*, 10 August 1872.
43 & 44 Victoria Cap. 42. *An Act to Extend and Regulate the Liability of Employers to make Compensation for Personal Injuries Suffered by Workmen in Their Service*, 7 September 1880.

Other published sources

Agricola, Georgius. *De Re Metallica* (New York: Dover Publications, 1950) trans. Herbert Clark Hoover and Lou Henry Hoover, http://www.gutenberg.org/files/38015/38015-h/38015-h.htm.
Anon. *The Lancashire Collier Girl. A True Story* (London: J. Marshall, 1795).
Anon. [Samuel Smiles], 'Workmen's Benefit Societies', *Quarterly Review*, 116:232 (October 1864), 318–50.
Arlidge, J. T. *The Hygiene, Diseases and Mortality of Occupations* (London: Percival and Co. 1892).
Bald, Robert. *A General View of the Coal Trade of Scotland, Chiefly That of the River Forth and Midlothian to Which Is Added, an Inquiry into the Condition of the Women Who Carry Coals Under Ground in Scotland, Known by the Name of Bearers, Etc.,* (Edinburgh: Oliphant, Waugh and Innes, 1812).
B., H. H. *Black Diamonds; or, the Gospel in a Colliery District* (London: James Herbert and Col, 1861).
Bristol Mining School. *Lectures Delivered at the Bristol Mining School, 1857* (Bristol: Bristol Mining School, 1859).
Buchan, William. *Domestic Medicine Or, a Treatise on the Prevention and Cure of Diseases by Regimen and Simple Medicines. With an Appendix, Containing a Dispensatory for the Use of Private Practitioners*. 11th Edition. (London: A. Strahan and others, 1790).
Buddle, John. *The First Report of a Society for Preventing Accidents in Coal Mines*. (Newcastle: Edward Walker, 1814).
Burt, Thomas, and Aaron Watson. *Thomas Burt, M.P., D.C.L., Pitman and Privy Councillor: an Autobiography, with Supplementary Chapters by Aaron Watson* (London: Unwin, 1924).
Campbell, George L. *Miners' Insurance Funds, their Origin and Extent* (London: Waterlow and Sons, 1880).
Dalziel, Alexander. *The Colliers' Strike in South Wales* (Cardiff: The Western Mail Offices, 1872).

Dodd, William. *A Narrative of the Experiences and Sufferings of William Dodd, A Factory Cripple*, 2nd edition (London: L. and G. Seeley, 1841).

Dodd, William. *The Factory System Illustrated in a Series of Letters to the Right Hon: Lord Ashley: Together with a Narrative of the Experience and Sufferings of William Dodd* (London: Frank Cass & Co., 1968).

Dunn, James. *From Coal Mine Upwards, Or, Seventy Years of an Eventful Life* (London: W. Green, 1910).

Eden, Frederick Morton. *Observations on Friendly Societies, for the Maintenance of the Industrious Classes, during Sickness, Infirmity, Old Age, and Other Exigencies* (London: J. White and J. White, 1801).

Engels, Frederick (sic). *The Condition of the Working-Class in England* (London: Panther, 1969).

Errington, Anthony [Edited by P. E. H. Hair]. *Coals on Rails, Or the Reason of My Wrighting: The Autobiography of Anthony Errington, a Tyneside Colliery Waggon and Waggonway Wright from his Birth in 1778 to around 1825* (Liverpool: Liverpool University Press, 1988).

Everett, James. *The Wall's End Miner, or, A Brief Memoir of the Life of William Crister Including an Account of the Catastrophe of June 18th, 1835* (London: Hamilton, Adams & Co., 1835).

Fynes, Richard. *The Miners of Northumberland and Durham: A History of Their Social and Political Progress* (Blyth: John Robinson, 1873).

Ginswick, Jules (ed.). *Labour and the Poor in England and Wales, 1849–1851: the Letters to the Morning Chronicle from the Correspondents in the Manufacturing and Mining Districts, the Towns of Liverpool and Birmingham, and the Rural Districts*, 8 vols. (London: Frank Cass, 1983).

Hurlstone, E. T. and J. P. Norman. *The Exchequer Reports: Reports of Cases Argued and Determined in the Courts of Exchequer & Exchequer Chamber*, vol. III (London: H. Sweet, W. Maxwell, and V. and R. Stevens and G. S. Norton, 1859).

Keelmen's Hospital Society. *Articles of the Keelmen's Hospital Society; with Rules and Regulations for the Hospital* (Newcastle upon Tyne: John Marshall, 1829).

Kentish, Edward. *An Essay on Burns, Principally upon Those Which Happen to Workmen in Mines* (London: G. G. and J. Robinson, 1797).

Kentish, Edward. *An Essay on Burns, In Two Parts: Principally on Those Which Happen to Workmen in Mines from Explosions of Carburetted Hydrogen Gas* (London: Longman et al. 1817).

Leifchild, J. R. *Our Coal and Our Coal-Pits* (London: Longman, Brown, Green, Longmans, 1855).

Mackenzie, E. *An Historical, Topographical, and Descriptive View of the County of Northumberland: And of Those Parts of the County of Durham Situated North of the River Tyne, with Berwick Upon Tweed, and Brief Notices of Celebrated Places on the Scottish Border* (Newcastle upon Tyne: Mackenzie and Dent, 1825).

Morgan, R. C. *The Life of Richard Weaver, the Converted Collier* (London: Morgan and Chase, 1861).

National Association of Coal, Lime and Iron-Stone Miners. *Transactions and Results of the National Association of Coal, Lime and Iron-Stone Miners of Great Britain, Held at Leeds, November 9, 10, 11, 12, 13 and 14 1863* (London: Longman and others, 1864).
Parkinson, George. *True Stories of Durham Pit-Life* (London: C. H. Kelly, 1912).
Ramazzini, Bernardino. *A Treatise of the Diseases of Tradesmen* (London: Andrew Bell, 1705).
Razzell, P. E. and R. W. Wainwright (eds), *The Victorian Working Class: Selections from Letters to the Morning Chronicle* (London: Frank Cass, 1973).
Rymer, Edward. *The Martyrdom of the Mine, or a Sixty Years Struggle for Life* (Middlesbrough, 1898), reprinted in Robert G. Neville (ed.), 'The Martyrdom of the Mine', *History Workshop* (Spring, 1976), 220–44 and (Autumn, 1976), 148–70.
Saunders, John. *Israel Mort, Overman* (London: Henry S. King, 1876).
Scrutator. *An Impartial Enquiry into the Existing Causes of Dispute between the Coal Owners of the Wear and Tyne and Their Late Pitmen* (Houghton-Le-Spring: Printed for the author by J. Beckwith, 1832).
Society for Bettering the Condition and Increasing the Comforts of the Poor. *The Reports of the Society for Bettering the Condition and Increasing the Comforts of the Poor*, vol. I (London: W. Bulmer and Co., 1798).
Symons, Jelinger C. *Tactics for the Times: As Regards the Condition and Treatment of the Dangerous Classes* (London: John Olliver, 1849).
Thackrah, Charles Turner. *The Effects of the Principal Arts, Trades and Professions and of Civic States and Habits of Living, on Health and Longevity* (London: Longman and others, 1831).
United Association of Colliers. *A Voice from the Coalmines, Or, A Plain Statement of the Various Grievances of the Pitmen of the Tyne and Wear: Addressed to the Coal Owners, their Head Agents, and a Sympathizing Public* (South Shields: J. Clark, 1825).
Weaver, Richard. *Richard Weaver's Life Story: the English Evangelist*, ed. James Patterson (Kilmarnock: J. Ritchie, 1897).
Wilson, John. *Autobiography of Ald. John Wilson, J.P., M.P … Reprinted from 'The Durham Chronicle'* (Durham: Durham Chronicle, 1909).
Wilson, Thomas. *The Pitman's Pay and other Poems* (Gateshead: William Douglas, 1842).

Online collections

The British Newspaper Archive, http://www.britishnewspaperarchive.co.uk/ .
Hansard, 1803–2005, http://hansard.millbanksystems.com/.
The Statistical Accounts of Scotland, 1791–1845, http://edina.ac.uk//stat-acc-scot.
Turner, David, Steven Thompson, Kirsti Bohata, Vicky Long, Arthur McIvor, Mike Mantin, Daniel Blackie, Ben Curtis, Angela Turner, Victoria Brown, Alexandra Jones, Anne Borsay. *Disability and Industrial Society, 1780–1948: A Comparative Cultural History of British Coalfields: Statistical Compendium*, http://doi.org/10.5281/zenodo.183686.

U.K. Parliamentary Papers (Proquest), http://parlipapers.proquest.com/.
Welsh Newspapers Online, http://newspapers.library.wales/.

Secondary sources

Published

Adams, Rachel, Benjamin Reiss and David Serlin (eds). *Keywords for Disability Studies* (New York: New York University Press, 2015).
Anderson, Julie. *War, Disability and Rehabilitation in Britain: Soul of a Nation* (Manchester: Manchester University Press, 2011).
Barnartt Sharon N. (ed.). *Disability as a Fluid State* (Bingley: Emerald, 2010).
Baron, Ava. 'Masculinity, the Embodied Male Worker, and the Historian's Gaze', *International Labor and Working-Class History*, 69 (2006), 143–60.
Barton, Len (ed.). *Disability and Society: Emerging Issues and Insights* (London: Longman, 1996).
Bartrip, P. W. J. *Workmen's Compensation in Twentieth-Century Britain: Law, History and Social Policy* (Aldershot: Gower, 1987).
Bartrip, P. W. J. *The Home Office and the Dangerous Trades: Regulating Occupational Disease in Victorian and Edwardian Britain* (Amsterdam: Rodopi, 2002).
Bartrip, P. W. J. and S. B. Burman. *The Wounded Soldiers of Industry: Industrial Compensation Policy 1833–1897* (Oxford: Clarendon Press, 1983).
Belchem, John. *Industrialization and the Working Class: The English Experience, 1750–1900* (Aldershot: Scolar Press, 1990).
Bengtsson, Steffan. 'Out of the Frame? Disability and the Body in the Writings of Karl Marx', *Scandinavian Journal of Disability Research*, 19:2 (2016), 151–60.
Benson, John. 'English Coal-Miners' Trade-Union Accident Funds, 1850–1900', *Economic History Review*, 28:3 (1975), 401–12.
Benson, John. 'Non-fatal Coalmining Accidents', *Bulletin of Society for the Study of Labour History*, 32 (1976), 20–2.
Benson, John. 'The Thrift of English Coal-Miners, 1860-95', *Economic History Review*, 31:3 (1978), 410–18.
Benson, John. *British Coalminers in the Nineteenth Century: A Social History* (Dublin: Gill and Macmillan, 1980).
Benson, John. 'Coalowners, Coalminers and Compulsion: Pit Clubs in England, 1860–80', *Business History*, 44:1 (2002), 47–60.
Benson, John. 'Coalminers, Coalowners and Collaboration: The Miners' Permanent Relief Fund Movement in England, 1860–1895', *Labour History Review*, 68:2 (2003), 181–94.
Berg, Maxine and Pat Hudson. 'Rehabilitating the Industrial Revolution', *Economic History Review*, 45:1 (1992), 24–50.
Berger, Stefan, Andy Croll, and Norman Laporte (eds). *Towards a Comparative History of Coalfield Societies* (Aldershot: Ashgate, 2005).
Beynon, Huw and Terry Austrin. *Masters and Servants: Class and Patronage in the*

Making of a Labour Organisation; the Durham Miners and the English Political Tradition (London: Rivers Oram, 1994).
Binding, Lucy. *The Representation of Bodily Pain in Nineteenth-Century English Culture* (Oxford: Oxford University Press, 2000).
Bohata, Kirsti, Alexandra Jones, Mike Mantin and Steven Thompson. *Disability in Industrial Britain: A Cultural History of Illness, Injury and Impairment in the Coal Industry, 1880-1948* (Manchester: Manchester University Press, Forthcoming).
Borsay, Anne. 'Returning Patients to the Community: Disability, Medicine and Economic Rationality before the Industrial Revolution', *Disability and Society*, 13 (1998), 645-63.
Borsay, Anne. *Disability and Social Policy in Britain Since 1750: A History of Exclusion* (Basingstoke: Palgrave Macmillan, 2005).
Borsay, Anne and Peter Shapely (eds). *Medicine, Charity and Mutual Aid: the Consumption of Health and Welfare in Britain c. 1550-1950* (Aldershot: Ashgate, 2007).
Bourke, Joanna. 'Housewifery in Working-Class England 1860-1914', *Past and Present*, 143 (1994), 167-97.
Bourke, Joanna. *What it Means to be Human: Reflections from 1791 to the Present* (London: Virago, 2011).
Bourke, Joanna. *The Story of Pain: From Prayer to Painkillers* (Oxford: Oxford University Press, 2014).
Brechin, Ann, Penny Liddiard and John Swain (eds). *Handicap in a Social World* (Sevenoaks: Hodder & Stoughton in association with the Open University Press, 1981).
Bronstein, Jamie L. *Caught in the Machinery: Workplace Accidents and Injured Workers in Nineteenth-Century Britain* (Stanford, CA: Stanford University Press, 2008).
Bronstein, Jamie L. 'The Hartley Colliery Disaster', *Victorian Review*, 40:2 (2014), 9-13.
Burch, Susan (ed.). *Encyclopedia of American Disability History* (New York: Facts on File, 2009).
Burch, Susan and Michael Rembis (eds). *Disability Histories* (Urbana, IL: University of Illinois Press, 2014).
Campbell, Alan B. *The Lanarkshire Miners: a Social History of their Trade Unions, 1775-1874* (Edinburgh: John Donald, 1979).
Challinor, Raymond and Brian Ripley. *The Miners' Association: A Trade Union in the Age of the Chartists* (London: Lawrence and Wishart, 1968).
Church, Roy. *The History of the British Coal Industry*, vol. 3: *1830-1913: Victorian Pre-eminence* (Oxford: Clarendon Press, 1986).
Colls, Robert. *The Collier's Rant: Song and Culture in the Industrial Village* (London: Croom Helm, 1977).
Colls, Robert. *Pitmen of the Northern Coalfield: Work, Culture, and Protest, 1790-1850* (Manchester: Manchester University Press, 1987).

Connop-Price, M. R. *Pembrokeshire: the Forgotten Coalfield* (Ashbourne: Landmark Publishing, 2004).

Conrad, Peter. *The Medicalization of Society: On the Transformation of Human Conditions into Treatable Disorders* (Baltimore, MD: Johns Hopkins University Press, 2007).

Cooter, Roger and Luckin, Bill (eds). *Accidents in History: Injuries, Fatalities and Social Relations* (Amsterdam: Rodopi, 1997).

Cooter, Roger and John Pickstone (eds). *Companion to Medicine in the Twentieth Century* (London and New York: Routledge, 2003).

Cordery, Simon. *British Friendly Societies, 1750–1914* (Basingstoke: Palgrave Macmillan, 2003).

Croll, Andy. 'Starving Strikers and the Limits of the "Humanitarian Discovery of Hunger" in Late Victorian Britain', *International Review of Social History*, 56 (2011), 103–31.

Croll, Andy. 'Strikers and the Right to Poor Relief in Victorian Britain: The Making of the Merthyr Tydfil Judgment of 1900', *Journal of British Studies*, 52 (2013), 128–52.

Crowther, M. A. 'Family Responsibility and State Responsibility in Britain before the Welfare State', *The Historical Journal*, 25 (1982), 131–45.

Curtis, Ben and Steven Thompson. '"A Plentiful Crop of Cripples Made by All This Progress": Disability, Artificial Limbs and Working-Class Mutualism in the South Wales Coalfield, 1890–1948', *Social History of Medicine*, 27 (2014), 708–27.

Curtis, Ben and Steven Thompson. '"This is the Country of Premature Old Men:" Ageing and Aged Miners in the South Wales Coalfield, c. 1880–1947', *Cultural and Social History*, 12:4 (2015), 587–606.

Curtis, Ben and Steven Thompson, 'Disability and the Family in South Wales Coalfield Society, c. 1920–1939', *Family & Community History* 20:1 (2017), 25–44.

Daunton, M. J. 'Miners' Houses: South Wales and the Great Northern Coalfield, 1880–1914', *International Review of Social History*, 25 (1980), 143–75.

Daunton, M. J. 'Down the Pit: Work in the Great Northern and South Wales Coalfields, 1870–1914', *Economic History Review*, 34:4 (1981), 578–97.

Davis, Lennard J. *Enforcing Normalcy: Disability, Deafness, and the Body* (London: Verso, 1995).

D'Cruze, Shani. *Crimes of Outrage: Sex, Violence and Victorian Working Women* (DeKalb: Northern Illinois University Press, 1998).

Deutsch, Helen and Felicity A. Nussbaum (eds). *'Defects': Engendering the Modern Body* (Ann Arbor, MI: University of Michigan Press, 2000).

De Veirman, Sofie. 'Deaf and Disabled? (Un)Employment of Deaf People in Belgium: a Comparison of Eighteenth-Century and Nineteenth-Century Cohorts', *Disability & Society*, 30 (2015), 460–74.

Digby, Anne. *The Evolution of British General Practice, 1850–1948* (Oxford: Oxford University Press, 1999).

Duckham, Helen and Baron, *Great Pit Disasters. Great Britain, 1700 to the Present Day* (Newton Abbott: David and Charles, 1973).

Duncan, Robert. *The Mineworkers* (Edinburgh: Birlinn, 2005).

Dupree, Marguerite W. 'Family Care and Hospital Care: the 'Sick Poor' in Nineteenth-Century Glasgow', *Social History of Medicine*, 6 (1993), 195–211.

Durbach, Nadja. *Spectacle of Deformity: Freak Show and Modern British Culture* (Berkeley CA: University of California Press, 2008).

Earwicker, Ray. 'Miners' Medical Services before the First World War: The South Wales Coalfield', *Llafur*, 3:2 (1981), 39–52.

Ernst, Waltraud (ed.). *Histories of the Normal and the Abnormal: Social and Cultural Histories of Norms and Normativity* (London and New York: Routledge, 2006).

Esmail, Jennifer and Christopher Keep. 'Victorian Disability: Introduction', *Victorian Review*, 35:2 (2009), 45–51.

Fewster, Joseph M. *The Keelmen of Tyneside: Labour Organisation and Conflict in the North-East Coal Industry* (Woodbridge: Boydell Press, 2011).

Finkelstein, Vic. *Attitudes and Disabled People: Issues for Discussion* (New York: World Rehabilitation Fund, 1980).

Flinn, Michael W. *The History of the British Coal Industry*, vol. 2: *1700–1830: The Industrial Revolution* (Oxford: Clarendon Press, 1984).

Foucault, Michel. *The Birth of the Clinic: An Archaeology of Medical Perception* (New York: Vintage Books, 1973).

Fox, Celina. 'The Development of Social Reportage in English Periodical Illustration During the 1840s and early 1850s', *Past and Present*, 74:1 (1977), 90–111.

Fraser, Derek (ed.). *The New Poor Law in the Nineteenth Century* (London: MacMillan, 1976).

Fryer, Jonathan. 'Preachers at the Pit: Methodists and County Durham Mining Disasters, 1880–1909', *Proceedings of the Wesley Historical Society*, 57:2 (2009), 25–31.

Garland Thomson, Rosemarie (ed.). *Freakery: Cultural Spectacles of the Extraordinary Body* (New York: New York University Press, 1996).

Gestrich, Andreas, Steven King and Lutz Raphael (eds). *Being Poor in Modern Europe: Historical Perspectives 1800–1940* (Oxford and Bern: Peter Lang, 2006).

Gier, Jaclyn J. and Laurie Mercier (eds). *Mining Women: Gender in the Development of a Global Industry* (New York: Palgrave Macmillan, 2006).

Gilbert, David. *Class, Community and Collective Action: Social Change in Two British Coalfields, 1850–1926* (Oxford: Oxford University Press, 1992).

Giles, Audrey C. 'Railway Accidents and Nineteenth-Century Legislation: "Misconduct, Want of Caution or Causes beyond their Control"', *Labour History Review*, 76:2 (2011), 121–42.

Gillis, John R. *For Better, for Worse: British Marriages, 1600 to the Present* (New York & Oxford: Oxford University Press, 1985).

Gleeson, Brendan J. *Geographies of Disability* (London: Routledge, 1999).

Gorsky, Martin. 'Mutual Aid and Civil Society: Friendly Societies in Nineteenth-Century Bristol', *Urban History*, 25 (1998), 302–22.

Gorsky, Martin. 'The Growth and Distribution of English Friendly Societies in the Early Nineteenth Century', *Economic History Review*, 51 (1998), 489–511.

Gorsky, Martin and Sally Sheard (eds), *Financing Medicine: the British Experience Since 1750* (Abingdon: Routledge, 2006).
Gray, Robert. 'Medical Men, Industrial Labour and the State in Britain, 1830–50', *Social History*, 16:1 (1991), 19–43.
Green, David G. *Working-Class Patients and the Medical Establishment: Self-Help in Britain from the Mid-Nineteenth Century to 1948* (Aldershot: Gower, 1985).
Griffin, Colin P. 'Methodism in the Leicestershire and South Derbyshire Coalfield in the Nineteenth Century', *Proceedings of the Wesley Historical Society*, 39 (1973–4), 62–72.
Griffin, Emma. *A Short History of the British Industrial Revolution* (Basingstoke: Palgrave Macmillan, 2006).
Griffin, Emma. *Liberty's Dawn: A People's History of the Industrial Revolution* (New Haven: Yale University Press, 2013).
Haines, Michael R. 'Fertility, Nuptiality, and Occupation: a Study of Coal Mining Populations and Regions in England and Wales in the Mid-Nineteenth Century', *Journal of Interdisciplinary History*, 8 (1977), 245–80.
Hair, P. E. H. 'Mortality from Violence in British Coal Mines, 1800–50', *Economic History Review*, 21:3 (1968), 545–61.
Harris, Bernard (ed.). *Welfare and Old Age in Europe and North America: The Development of Social Insurance* (London: Pickering & Chatto, 2012).
Harrison, Royden (ed.). *Independent Collier: The Coal Miner as Archetypal Proletarian Reconsidered* (New York: St Martin's Press, 1978).
Hassan, John A. 'The Landed Estate, Paternalism and the Coal Industry in Midlothian, 1800–1880', *The Scottish Historical Review*, 59:167 (1980), 73–91.
Heesom, A. J. 'Entrepreneurial Paternalism: The Third Lord Londonderry (1778–1854) and the Coal Trade', *Durham University Journal*, 35 (1974), 238–56.
Hirsch, Jerrold and Karen Hirsch. 'Disability in the Family?: New Questions About the Southern Mill Village', *Journal of Social History*, 35:4 (Summer, 2002), 919–33.
Holdren, Nate. 'Incentivizing Safety and Discrimination: Employment Risks under Workmen's Compensation in the Early Twentieth Century United States', *Enterprise and Society*, 15 (2014), 31–67.
Holmes, Martha Stoddard. *Fictions of Affliction: Physical Disability in Victorian Culture* (Ann Arbor: University of Michigan Press, 2004).
Hopkins, Eric. *Working Class Self-Help in Nineteenth-Century England: Responses to Industrialisation* (London: UCL Press, 1995).
Humphries, Jane. *Childhood and Child Labour in the British Industrial Revolution*. (Cambridge: Cambridge University Press, 2010).
Hutchison, Iain. *A History of Disability in Nineteenth-Century Scotland* (Lewiston, NY: Edwin Mellen Press, 2007).
Jackson, Mark (ed.). *The Routledge History of Disease* (London: Routledge, 2017).
Jewson, N. D. 'The Disappearance of the Sick-Man from Medical Cosmology, 1770–1870', *Sociology*, 10 (1976), 225–44.

John, Angela V. *By the Sweat of Their Brow: Women Workers at Victorian Coal Mines* (London: Routledge & Kegan Paul, 1984).
John, Angela V. (ed.). *Our Mothers' Land: Chapters in Welsh Women's History, 1830–1939* (Cardiff: University of Wales Press, 1991).
Jones, David J. V. *Before Rebecca: Popular Protests in Wales, 1793–1835* (London: Allen Lane, 1973).
Jones, Dot. 'Did Friendly Societies Matter? A Study of Friendly Society Membership in Glamorgan, 1794–1910', *Welsh History Review*, 12:3 (1985), 324–49.
King, Steven. *Poverty and Welfare in England 1700–1850: A Regional Perspective* (Manchester: Manchester University Press, 2000).
King, Steven. '"Stop This Overwhelming Torment of Destiny": Negotiating Financial Aid at Times of Sickness under the English Old Poor Law, 1800–1840', *Bulletin of the History of Medicine*, 79:2 (2005), 228–60.
King, Steven. 'Constructing the Disabled Child in England, 1800–1860', *Family & Community History*, 18 (2015), 104–21.
King, Steven, and Geoffrey Timmins. *Making Sense of the Industrial Revolution: English Economy and Society, 1700–1850* (Manchester: Manchester University Press, 2001).
Kirby, Peter. *Child Workers and Industrial Health in Britain, 1780–1850* (Woodbridge: Boydell Press, 2013).
Kirkup, John. *A History of Limb Amputation* (London: Springer, 2007).
Kudlick, Catherine J. 'Disability History: Why We Need Another "Other"', *American Historical Review*, 108 (2003), 763–93.
Kudlick, Catherine J. 'Comment: On the Borderlands of Medical and Disability History', *Bulletin of the History of Medicine*, 87:4 (2013), 540–59.
Lawrence, Christopher. *Medicine in the Making of Modern Britain 1700–1920* (London and New York: Routledge, 1994).
Lee, Robert. *The Church of England and the Durham Coalfield 1810–1926: Clergymen, Capitalists, and Colliers* (Woodbridge: Boydell Press, 2007).
Leneman, Leah. 'Lives and Limbs: Company Records as a Source for the History of Industrial Injuries', *Social History of Medicine*, 6:3 (1993), 405–27.
Levine-Clark, Marjorie. '"The Entombment of Thomas Shaw:" Mining Accidents and the Politics of Workers' Bodies', *Victorian Review*, 40:2 (2014), 22–6.
Linker, Beth. 'On the Borderland of Medical and Disability History: A Survey of the Fields', *Bulletin of the History of Medicine*, 87:4 (2013), 499–535.
Longmore, Paul K. and Lauri Umansky (eds). *The New Disability History: American Perspectives* (New York: New York University Press, 2001).
Longmore, Paul K. *Why I Burned My Book and Other Essays on Disability* (Philadelphia: Temple University Press, 2003).
Luckin, Bill and Roger Cooter (eds). *Accidents in History: Injuries, Fatalities and Social Relations* (Amsterdam and Atlanta GA: Rodopi, 1997).
McIvor, Arthur and Ronald Johnston. *Miners' Lung: A History of Dust Disease in British Coal Mining* (Aldershot: Ashgate, 2007).

McRuer, Robert and Anna Mollow (eds). *Sex and Disability* (Durham, NC.: Duke University Press, 2012).

Mantin, Mike. 'Coalmining and the National Scheme for Disabled Ex-Servicemen after the First World War', *Social History*, 41:2 (2016), 155–70.

Marshall, J. D. *The Old Poor Law, 1795–1834* (Second Edition. Basingstoke: Macmillan, 1985).

Metcalfe, Alan. *Leisure and Recreation in a Victorian Mining Community: the Social Economy of Leisure in North-East England, 1820–1914* (London: Routledge, 2006).

Mills, Catherine. *Regulating Health and Safety in the British Mining Industries, 1800–1914* (Farnham: Ashgate, 2010).

Mitchell, David T. and Sharon Snyder. *Narrative Prosthesis: Disability and the Dependencies of Discourse* (Ann Arbor MI: University of Michigan Press, 2000).

Mokyr, Joel (ed.). *The Oxford Encyclopedia of Economic History*, Vol. 5 (Oxford & New York: Oxford University Press, 2003).

Moore, R. I. *Pit-Men, Preachers and Politics: the Effects of Methodism in a Durham Mining Community* (Cambridge: Cambridge University Press, 1974).

Morris, J. H. and L. J. Williams. *The South Wales Coal Industry, 1841–1875* (Cardiff: University of Wales Press, 1958).

Mounsey, Chris (ed.). *The Idea of Disability in the Eighteenth Century* (Lewisburg PA: Bucknell University Press, 2014).

Nielsen, Kim E. *A Disability History of the United States* (Boston: Beacon, 2012).

O'Leary, Paul. *Immigration and Integration: The Irish in Wales 1798–1922* (Cardiff: University of Wales, 2000).

Oliver, Michael and Colin Barnes. *The New Politics of Disablement* (Basingstoke: Palgrave Macmillan, 2012).

Pickstone, John V. *Medicine and Industrial Society: A History of Hospital Development in Manchester and its Region 1752–1946* (Manchester: Manchester University Press, 1985).

Price, Kim. *Medical Negligence in Victorian Britain: the Crisis of Care under the English Poor Law* (London: Bloomsbury Academic, 2015).

Quick, Michael. 'Mid-Victorian Compensation Culture', *Journal of the Railway and Canal Historical Society*, 192 (2005), 110–17.

Quinn, Michael. 'Jeremy Bentham on Physical Disability: A Problem for Whom?', *Review of Disability Studies: An International Journal*, 8:4 (2012), 19–32.

Rembis, Michael, Catherine Kudlick and Kim E. Nielsen (eds). *The Oxford Handbook of Disability History* (New York: Oxford University Press, Forthcoming).

Richardson, Ruth. *Death, Dissection and the Destitute* (Chicago and London: Chicago University Press, 2000).

Riley, James C. *Sick, Not Dead: The Health of British Workingmen during the Mortality Decline* (Baltimore, MD: Johns Hopkins University Press, 1997).

Rose, Sarah F. '"Crippled" Hands: Disability in Labor and Working-Class History', *Labor* 2:1 (2005), 27–54.

Rose, Sarah F. *No Right To Be Idle: The Invention Of Disability, 1840s–1930s* (Chapel Hill: University of North Carolina Press, 2017).

Rosen, George. *The History of Miners' Diseases: A Medical and Social Interpretation*, with an Introduction by Henry E. Sigerist (New York: Schuman's, 1943).

Rule, John. *The Labouring Classes in Early Industrial England, 1750–1850* (London: Longman, 1986).

Sandahl, C. and P. Auslander (eds). *Bodies in Commotion: Disability and Performance* (Ann Arbor: University of Michigan Press, 2005).

Sharpe, Pamela. 'Explaining the Short Stature of the Poor: Chronic Childhood Disease and Growth in Nineteenth-Century England', *Economic History Review* 65:4 (2012), 1475–94.

Slavishak, Edward. *Bodies of Work: Civic Display and Labor in Industrial Pittsburgh* (Durham NC and London: Duke University Press, 2008).

Smith, F. B. *The People's Health, 1830–1910* (London: Croom Helm, 1979).

Stewart, John and Steve King. 'Death in Llantrisant: Henry Williams and the New Poor Law in Wales', *Rural History* 15:1 (2004), 69–87.

Stiker, Henri-Jacques. *A History of Disability* (Ann Arbor: University of Michigan Press, 1999).

Stone, Deborah. *Disabled State* (London: Macmillan, 1985).

Strange, Julie Marie. *Fatherhood and the British Working Class, 1865–1914* (Cambridge: Cambridge University Press, 2015).

Szabo, Jason. *Incurable and Intolerable: Chronic Disease and Slow Death in Nineteenth-Century France* (New Brunswick NJ: Rutgers University Press, 2009).

Szreter, Simon. *Fertility, Class and Gender in Britain, 1860–1940* (Cambridge: Cambridge University Press, 1996).

Thompson, E. P. 'Time, Work-Discipline and Industrial Capitalism', *Past and Present*, 38 (1967), 56–97.

Thompson, E. P. *The Making of the English Working Class* (Harmondsworth: Penguin, 1976).

Tonks, David. 'A Kind of Life Insurance: the Coal-Miners of North-East England 1860–1920', *Family and Community History*, 2:1 (1999), 45–58.

Turner, David M. *Disability in Eighteenth-Century England: Imagining Physical Impairment* (New York: Routledge, 2012).

Turner, David M. '"Fraudulent" Disability in Historical Perspective', *History and Policy* (14 February 2012) http://www.historyandpolicy.org/policy-papers/papers/fraudulent-disability-in-historical-perspective.

Watson, Nick, Alan Roulstone and Carol Thomas (eds). *Routledge Handbook of Disability Studies* (London: Routledge, 2012).

Weindling, Paul (ed.). *The Social History of Occupational Health* (London: Croom Helm, 1985).

Whatley, Christopher A. 'A Caste Apart? Scottish Colliers, Work, Community and Culture in the Era of "Serfdom", c. 1606–1799', *Journal of the Scottish Labour History Society*, 26 (1991), 3–20.

Williams, John. *Was Wales Industrialised?: Essays in Modern Welsh History* (Llandysul, Dyfed: Gomer Press, 1995).
Williams-Searle, John. 'Courting Risk: Disability, Masculinity, and Liability on Iowa's Railroads, 1868–1910', *The Annals of Iowa*, 58 (1999), 27–77.
Wilson, James C. and Cynthia Lewiecki-Wilson (eds). *Embodied Rhetorics: Disability in Language and Culture* (Carbondale and Edwardsville: Southern Illinois Press, 2001).
Wohl, Anthony S. *Endangered Lives: Public Health in Victorian Britain* (London: J. M. Dent, 1983).
Wrigley, E. A. *Continuity, Chance and Change: the Character of the Industrial Revolution in England* (Cambridge: Cambridge University Press, 1990).

Unpublished

Blackie, Daniel. 'Disabled Revolutionary War Veterans and the Construction of Disability in the Early United States, c. 1776–1840' (PhD thesis, University of Helsinki, 2010), http://urn.fi/URN:ISBN:978-952-10-6343-5.
Howard, William Stuart. 'Miner's Autobiographies, 1790–1945: A Study of Life Accounts by English Miners and their Families' (PhD thesis, Sunderland Polytechnic, 1991).
Jones, Alexandra. 'Disability in Coalfields Literature c. 1880–1948: A Comparative Study' (PhD thesis, Swansea University, 2016).
Lewis, Halle Gayle. '"Cripples are not the Dependents One is Led to Think": Work and Disability in Industrializing Cleveland, 1861–1916' (PhD thesis, State University of New York at Binghamton 2004).

Index

accidents
  documentation of 165, 190
  fatal 1, 3, 30, 129, 134, 165
  non-fatal 1–2, 3, 23, 26, 31–6, 47, 64, 66, 67, 105, 165
age
  and earnings 46
  and exposure to risk 35, 46
  and status 28, 39, 48, 201
  old 46, 98, 99
Alison, Somerville Scott 37, 58, 63, 74, 98, 102, 143
amputation 35, 39, 59, 74, 114, 129, 152
  as spiritual trial 151
  resistance to 78–9, 204
Arlidge, J. T. 40, 56, 57

Bald, Robert 60
Barnes, Colin 4, 5
Baron, Ava 201
Bartrip P. W. J. 185, 189
begging 135
Benson, John 1, 27, 167
'Blind Miner of Bottalack', the 117
blindness 38, 73, 117
  see also eyes and eyesight
'bodily capital' 165
boils 40, 63
'bord and pillar' mining 42–3
Borsay, Anne 6
Bourke, Joanna 59
'boys' work' 39, 48, 144, 201
Bristol Mining School 65, 67, 69, 117
British Miners' Benefit Association 116
Bronstein, Jamie 35, 177, 184
  Caught in the Machinery (2008) 13
Brough, Lionel 35
Buchan, William 69
Buddle, John 32, 34, 38
Bullock, Jim 134, 148
Bunyan, John
  Pilgrim's Progress (1678) 150
Burman, S. B. 185, 189
burns 1, 35, 38, 60, 68, 70, 71, 72, 75, 114

Burt, Thomas 23, 28, 33, 34, 37, 72, 75, 188

Carluke (Lanarkshire) 102, 103, 138
cavilling 43–4, 48
census 138, 145
Chadwick, Edwin 57
charity 106, 114, 144
Chatham Chest 100
Chicken, Edward
  The Collier's Wedding (1729) 130
childbirth 63, 154
children
  employment of 25, 28, 37, 63, 95–6, 142
Children's Employment Commission (1842) 2, 24, 25, 37, 39, 42, 45, 57, 58, 59–60, 62–4, 65, 66, 70, 79, 82, 95, 96, 101, 111, 163, 169–70, 203, 205
cholera epidemic (1831–32) 167
Church, Roy, 95
class struggle
  and disabled people 183
coal
  and industrial growth 3, 8–12, 30, 32, 33, 46
coal industry
  employment in 9
  output 9–10
  see also women, employment in coal industry
coal mines
  inspection of, 12, 16, 31–2, 33, 36, 47, 164–5
coalfields
  communities 3, 11, 83, 93, 100, 175, 177, 178, 182–3, 202, 203
  geological differences between 11, 24, 26, 33, 44, 56
  Great Northern (north east England) 3, 11, 25, 27, 29, 30, 31, 32, 34, 35, 40, 41, 42, 45, 57, 65, 70, 96, 98, 99–100, 101, 171, 174, 175, 180

Lancashire 24, 80, 171
  migration into 130
  Scotland 11, 25, 30, 32, 34, 36, 45,
    61–2, 96, 107, 180, 184
  south Wales 11, 24, 27, 28, 30, 31,33,
    35, 36, 42, 45, 56, 65, 66, 72, 74, 76,
    80, 118, 177, 191
  variations in employment practices
    between 27–30, 33, 43, 47
  Yorkshire 24, 134, 171
coalminers
  character of 128–30
  healthiness of 56–7
  'heroic self-dependence' of 93–4, 100,
    117
  intellectual characteristics of 58
  as a 'race of men' 58–9, 139, 204
  reputation for physical strength 55,
    57
  stereotypes of 130–1, 154
coalmining occupations 23, 25
  ancillary work 27–8
  hewing 25, 26–9, 38, 42, 44, 45, 96,
    144
  hierarchy of occupations 26, 28
  overmen 40, 111
  putting 28, 35, 38
collieries
  Backworth 115
  Beamish 108
  Birsley 37
  Blaydon Main 96
  Carfin 108, 112
  Collinshiel 61
  Cramlington 23, 182
  Cymllynfell 111
  Dry Clough 45
  Gosforth 40
  Govan 179–80
  Graig 111
  Hallbeath 62
  Hebburn 111
  Hetton 62, 179
  Houghton 44
  Jarrow 134, 149
  Loughor 66
  Middle Duffryn 80
  Monkwearmouth 40, 64, 65
  Murton 23, 34, 75

Nantyglo 81
Newbottle 79
Old Grange 44
Pease Deanery 65
Penston 95
Percy Main 37
Polkemmet 37
Risca 66, 70
Seaton Delavel 28
South Dunraven 81
South Hetton 25
Thornley 40
Tyne Main 39
Walbottle 114
Walker 60
Washington 171
Waterloo 42
Willington 66
Wylam 56
Ynyscynon 97
see also disasters, mining
Colliery Guardian 26, 171, 184, 187
Colls, Robert 148, 174, 176
compensation 12, 32, 80, 183–92, 203
  and 'fellow servant' rule 183, 185
  and litigation 80, 183–5
contracts 29–30, 38, 98
Cordery, Simon 110
Cossham, Handel 117
Cowen, Joseph 16, 187–8
crime 97, 131–2, 143
crushing 16, 34, 35, 78, 114
Curtis, Ben 191

Davis, Lennard 14
Davy, Humphrey 34
  see also safety lamp, miner's
D'Cruze, Shani 132
de la Beche, Henry 2, 15
De Veirman, Sophie 7
'deadwork' 27
Defoe, Daniel 100
deformity 36, 40, 57, 58, 82, 204
disability
  and critique of industrial capitalism
    164, 166, 169–70
  definitions of 13–15, 35, 36–7, 101,
    102–3, 113, 119
  and disease 60

and educational opportunities 143, 203
and emotions 141–2, 182
and gendered identities 145–6, 154
historical materialist approaches to 4–6, 36, 200
and industrial relations 165
and life cycle 146
medical model of 77
neglect in labour history 8
as permanent condition 119, 204
and physical appearance 147–8
and prejudice, ridicule, and discrimination 135–6, 192, 201
psychological consequences of 133
and race 59
and religious activism 153
and reproductive health 63
and sex 145, 202
and suffering 170, 180–1, 204–5
and 'variability' 15
and work 36–46, 63, 175, 200–2
disasters, mining 1, 30, 169, 200
  Abercarn (1878) 98, 106
  Blantyre Colliery (1877) 30
  Gethin Colliery (1862) 42
  Gethin Colliery (1865) 40, 43, 93, 98, 140
  Hartley Colliery (1862) 1, 94, 114, 149, 184
  Haswell Colliery (1844) 174
  Jarrow Colliery (1826) 134, 149
  Universal Colliery, Senghenydd (1913) 1
  Wallsend (1835) 71
  see also accidents, fatal; explosions
diseases of coalminers 59–64
disfigurement 36, 58
dissection 61–2, 203
doctors
  colliery 25, 31, 64–7, 69, 70, 77–82, 83
Dodd, William 7
Dunn, Matthias 68

Easington (County Durham) 9, 98, 101
Eden, Frederick Morton 108

Edinburgh 61, 62
Engels, Friedrich 7, 169–70
  Condition of the Working Class in England (1844) 169
epilepsy 102
Errington, Anthony 70
ethnicity 130, 135
eugenics 7, 14
Everett, James 71
explosions 1, 30–1, 32, 33, 34, 42–3, 93, 174
eyes and eyesight 34, 36, 38, 58, 73–4, 144, 147

factories 6–7, 12, 37
Factory Commission (1833) 7
family 42–3, 76, 138–9, 141–2, 154, 201–2
  role in social welfare 95–8
  see also home; households
Finkelstein, Vic 4, 5
first aid 69
fractures 35, 68, 73, 75, 80
Franks, Robert 63, 96
fraud 108, 117, 135, 187
friendly societies 16, 65, 75, 79, 83, 93, 105, 106–13, 115, 117, 118, 167, 202, 204
  affiliated orders 109
  and industrial politics 178–9
  membership of 107, 109–10
  payments to members 112–13, 115
  rules 110–11
Fynes, Richard, 175, 180, 182

Garland Thomson, Rosemarie 5
Gisborne, Thomas 29
Glasgow 61, 73, 74, 104
Gleeson, Brendan 4, 5
  see also somatic flexibility
Gregory, James 133

healers, unorthodox 70–1
heat
  in coal mines 26, 33, 40, 55, 63
heredity 58
Holmes, Martha Stoddard 37

home 136–48
    as location of care 72–3
    and gendered division of labour 138–41
hospitals 12, 73–5, 82
    Glasgow Ophthalmic Institution 73
    Glasgow Royal Infirmary 73, 74
    Guy's 62
    Keelmen's 73, 75, 82
    Manchester Infirmary 146, 151
households
    composition of 138–9
    and householder status (headship) 143, 145–6, 155
    lodgers 140
    see also home
houses, company owned 97, 140
    evictions from 180–2, 205
humoralism 62, 70
Humphries, Jane 139, 142, 143
Hutchison, Iain 103
hygiene 57, 64

Illustrated London News 136, 180
Industrial Revolution
    historical approaches to 6
    in disability studies 4–5, 173
industrialisation 2–3, 6, 11, 163
    'industrialisation thesis' in disability studies 4–8, 23–4, 41
institutionalisation 5
Irish 130, 135, 175
iron industry 29, 45, 47, 81, 128

keelmen 73, 187
Kentish, Edward 1, 65, 71, 72
    Essay on Burns (1797) 60, 79
King, Steven 105
Kirby, Peter 7

lameness 37, 38, 39, 40, 99, 114
Larkhall Miner's Mutual Protection, Accident and Funeral Association 167
Leeds 57
legislation
    Employers' Liability Act (1880) 4, 183, 186–9, 191

Act to Consolidate and Amend the Acts Relating to the Regulation of Coal Mines and Certain Other Mines [Coal Mines Act] (1872) 31, 186
Act for Inspection of Coal Mines in Great Britain [Mines Inspection Act] (1850) 31, 169, 183
Act to Prohibit the Employment of Women and Girls in Mines and Collieries, to Regulate the Employment of Boys, and to Make other Provisions Relating to Persons Working Therein [Mines and Collieries Act] (1842) 25, 28, 96, 164, 169
Medical Act (1858) 70
Poor Law Amendment Act (1834) 101
Truck Act (1831) 66
Workmen's Compensation Act (1897) 191
Leifchild, John Roby 31, 41, 58, 100
life insurance 113–14
Longmore, Paul 77
longwall mining 45–6, 47
lung diseases 12, 33, 36, 60–2, 64, 77, 96, 98, 104, 105, 138, 141–2, 200

Macdonald, Alexander 186–8
Mackworth, Herbert 36, 65, 67, 69
marriage 139–40, 145–8, 155, 202
Marx, Karl 7
masculinity 13, 25, 40, 47–8, 107, 132, 134, 143–5, 154–5, 202–3
    and breadwinner role 107, 132, 139, 140, 142, 143, 145, 146, 154, 202
    and trade union membership 170
mechanisation 5, 46
medical authority 61–2, 70, 77–82, 83, 203–4
medical research 61–2
mental health 14, 16, 71
Merthyr Tydfil 41, 42, 57, 66, 81, 98, 101, 106, 111, 112, 129, 131, 150, 177
Mills, Catherine 165
Miner and Workmen's Advocate 55, 70, 78, 81, 113, 116
Mitchell, Dr James 1, 15, 31

morbidity
  in coalmining 26, 31–6
Morning Chronicle 129, 150
mortality
  in coalmining 30, 33, 56, 165, 190
Mounsey, Chris 16

'narrative prosthesis' 170
Newcastle-upon-Tyne 1, 16, 60, 66, 67, 70, 73, 75

occupational health
  history of 12–13
Oliver, Michael 4, 5

pain 37, 58–9, 71
Parkinson, George 26, 42, 151
Parliamentary Select Committees
  on Accidents in Coalmines (1835) 28
  on Medical Poor Relief (1844) 100
  on Mines (1866) 102
paternalism 60, 65–7, 83, 94, 99–100, 118, 165, 182, 187, 203, 206
Penny Magazine 9, 25, 26, 57
Permanent Relief Fund,
  Northumberland and Durham
  Miners' 17, 94, 114–17, 118, 188
piecework 27, 42, 201
pit clubs (*see* sick clubs)
Pontypool 57, 75, 108, 112
Poor Law
  England and Wales 14, 16, 64, 65, 94, 98, 99, 100, 101–2, 105, 118, 144, 170, 177
  medical services 74–5, 100
  Scotland 96, 102–5, 138
posture 26–7, 59–60
poundage 66, 81
prosthetics 23, 39, 43, 75–6, 106, 131, 132
Punch 166

Ramazzini, Benardino
  De Morbis Artiricium Diatriba (1700) 59
recklessness 23, 34–5, 78, 129, 172
rehabilitation 66, 75–6, 83
religion
  Anglican Church 149
  and bodily suffering 150–1
  chapels 148–9
  and charity 149
  Methodism 129, 148–52
  and power of prayer 149–50, 153
  sectarianism 149
  and spiritual comfort 149
rescue 67–8
Rest, the (Porthcawl) 76, 83
rheumatism 38, 63, 70
Riley, James C. 78
roof falls 27, 33, 35, 36, 39, 46
Rose, Sarah 8, 165, 170
Rosen, George
  The Diseases of Miners (1943) 12
Royal Humane Society 69
Rymer, Edward 38, 39, 44, 48, 97, 143, 147, 171, 173, 177, 191, 204

St John Ambulance movement 68, 69
safety 31–4, 45, 64, 168–9, 190, 203
safety lamp, miner's 34, 39
sanitation 57, 60, 72
Saunders, John
  Israel Mort, Overman (1876) 142
scars 40, 47
schools 41, 66
'Scotch Cattle' 173, 175
seasoning 39
serfdom
  in Scottish coalmining 25, 30
settlement 73, 99
sick clubs 55, 64–7, 81–2, 83, 93, 94, 108, 117, 167, 203
Simpson, Dr James Y. 61
Slavishack, Edward 39
smart money 99–100, 167–8, 177–8, 185, 187, 190
Smiles, Samuel 106
Society for the Prevention of Accidents
  in Coalmines 32
somatic flexibility 5, 36, 41, 45, 176, 190–1, 201, 203
Stewart, Patrick 97
stigma 39
Stone, Deborah 14
Stonelake, Edmund 70, 74, 107

strikes 40, 116, 166, 169, 175, 177–80, 182–3
  as cause of incapacity and injury 172–4
  and strike-breaking 175, 180
superstition 70, 80
surgeons 56, 57, 61–2, 65–7, 69, 70, 79–81, 100, 111, 203
surveillance
  medical 111–12, 117
Symons, Jelinger C.
  Tactics for the Times (1849) 128

teaching
  as occupation for disabled miners 41
Thackrah, Charles Turner 59
  The Effects of the Principal Arts, Trades and Professions ... on Health and Longevity (1831) 57
Thain, James 45, 48
Thompson, Steven 191
Thomson, J. B. 64
Thomson, Dr William 61–2
timekeeping 5, 44
Tissot, Samuel 69
Towers, John 116
trade unions 12, 30, 34, 93, 113, 116, 166–72, 178, 184, 186, 190, 204
  Durham Miners' Association 167, 186, 188, 190
  membership of among mineworkers 171
  National Association of Coal Miners 81
  and restriction of output 176, 201
  United Association of Colliers 168
  see also Trades Union Congress (TUC)

Trades Union Congress (TUC) 186
Tranent (East Lothian) 37, 72, 74, 98, 102, 111, 113, 143
transportation
  of wounded 67–9
Tremenheere, Hugh Seymour 164, 169, 173, 176

ventilation 33, 63, 168–9
violence 110, 131–2, 143, 172–3

wages and earnings 27, 29, 35, 42, 44, 66, 106
  as compensation for risk 29, 185
washing 57, 154
  see also hygiene
Weaver, Richard 129, 146, 151, 153
Weindling, Paul 12
welfare 203–4
  mixed economy of 94, 117
  and strikes 177
  see also friendly societies; Poor Law
Williams, John 42
Willams-Searle, John 172
Wilson, Dr Robert 109
Wilson, John 72, 75, 135
Wilson, Thomas
  The Pitman's Pay (1843) 130
Wohl, Anthony S. 56
women
  and domestic work 140–1, 144, 205
  employment in coal industry 24–5, 42, 63, 95–6, 139, 205
  health affected by work 63, 141, 154
  see also childbirth
workhouses 74–5, 101, 103–4, 105
Wrigley, E. A. 8

EU authorised representative for GPSR:
Easy Access System Europe, Mustamäe tee 50,
10621 Tallinn, Estonia
gpsr.requests@easproject.com

www.ingramcontent.com/pod-product-compliance
Ingram Content Group UK Ltd.
Pitfield, Milton Keynes, MK11 3LW, UK
UKHW021128160426
5217IPUK00046B/68